A TREATISE ON INDUCTION
AND PROBABILITY

Founded by C. K. Ogden

The International Library of Philosophy

PHILOSOPHY OF LOGIC AND MATHEMATICS
In 8 Volumes

A TREATISE ON INDUCTION
AND PROBABILITY

GEORG HENRIK VON WRIGHT

First published in 1951 by
Routledge and Kegan Paul Ltd

Reprinted in 2000, 2001 (twice) by
Routledge

2 Park Square, Milton Park, Abingdon, Oxfordshire, OX14 4RN
270 Madison Avenue, New York NY 10016

Routledge is an imprint of the Taylor & Francis Group

First issued in paperback 2010

The publishers have made every effort to contact authors/copyright holders
of the works reprinted in the *International Library of Philosophy*.
This has not been possible in every case, however, and we would
welcome correspondence from those individuals/companies
we have been unable to trace.

These reprints are taken from original copies of each book. In many cases
the condition of these originals is not perfect. The publisher has gone to
great lengths to ensure the quality of these reprints, but wishes to point
out that certain characteristics of the original copies will, of necessity, be
apparent in reprints thereof.

British Library Cataloguing in Publication Data
A CIP catalogue record for this book
is available from the British Library

A Treatise on Induction and Probability
ISBN 978-0-415-22548-5 (hbk)
ISBN 978-0-415-61362-0 (pbk)
Philosophy of Logic and Mathematics: 8 Volumes
ISBN 978-0-415-22575-5
The International Library of Philosophy: 56 Volumes
ISBN 978-0-415-21803-0

To C.D.B.

CONTENTS

7

CONTENTS

CHAPTER FIVE

INDUCTION AND DEDUCTION

CHAPTER SIX

INDUCTION AND DEFINITION

CHAPTER SEVEN

THE LOGIC OF PROBABILITY

CONTENTS

CHAPTER EIGHT

PROBABILITY AND PREDICTION

CHAPTER NINE

PROBABILITY AND LAWS OF NATURE

9

CONTENTS

CHAPTER TEN

INDUCTION AND INVERSE PROBABILITY

PREFACE

THE present book is primarily a treatise on induction. As such its aim is to examine, in the light of standards of logical correctness, various types of argument which can be grouped under the common heading of induction. I shall sometimes talk of this examination as a reconstruction of arguments.

Treatment of induction is intimately connected with treatment of probability. On the foundations of probability there has been an extensive literature within recent years. So far, however, no altogether satisfactory basis has been provided on which to accomplish the tasks set forth in this book relating to so-called Inductive Probability. It has therefore been necessary to devote considerable space to a discussion of probability in general.

Considering the importance of the subject, the number of books which deal with induction is extraordinarily small. This observation was made by the late Lord Keynes in his admirable *Treatise on Probability*. It remains substantially true even to-day. What has been written on induction since the appearance of Keynes' book in 1921 has, with few exceptions, been confined either to a description and criticism of Mill's methods, or to a discussion of the Problem of Hume.

In the last few years, however, a new interest in the subject seems to have awoken, chiefly among authors influenced by modern logic. It was, in fact, to be expected that the rapid development of logic in our times, having long since found fruitful applications in the fields of mathematics and the exact sciences, should at last become felt also in the backward realm of induction.

The hope of contributing to the opening of a new era in the history of the subject, in which the topic of induction shall no longer deserve the name of a scandal to philosophy, has encouraged me to publish the present book. It is, however, only after great hesitation that I submit my work to the public. The more insight I have acquired into the subject-matter of my investigation, the more I have been compelled to narrow the range and relevance of my exposition in order to secure it upon a solid foundation. Yet most of what is said here is bound to be of a merely provisional and relative nature. I know that the contribution of this book to the subject is but a feeble beginning, in need of completion and extension in several directions.

A main defect of most logical investigations into the nature of scientific thinking is that of oversimplification and of the assimilation of cases to general patterns which they do not really fit. This defect is due to two reasons: failure to state arguments with sufficient precision and lack of insight into the complex nature of scientific thought and practice. In this book I have been mainly concerned with making arguments clear. The problem to what extent the arguments have the power to illuminate by reconstruction the actual procedures of science is, on the whole, not discussed. This onesidedness may perhaps create the impression that the author claims for abstract ideas a relevance and applicability which they do not possess. I hope to be able later to compensate for this narrowness, and I am anxious to dissociate myself from any claims to an all-embracing and unrestricted applicability of my ideas for the purpose of understanding the nature of scientific thought.

The author from whom I have learnt most is undoubtedly Keynes. It seems to me that next to Francis Bacon, his has been the most fertile mind seriously to occupy itself with the questions which are the main topic of this inquiry. I have also drawn much inspiration from the works of von Mises and Reichenbach on probability.

During the course of my studies I have been very much indebted to my friend, Professor C. D. Broad of Trinity College, Cambridge. Already as a research student in England before the war I had the privilege of having detailed discussions with him on the topics of induction and probability. He has also been good enough to read the manuscript of the present book and he has contributed to the development of my thoughts with numerous valuable suggestions. Without his unselfish assistance and interest this monographic exposition of my ideas would probably not have been written.

Cambridge,   G. H. von Wright.
December, 1948.

Chapter One

INDUCTION AND ITS PROBLEMS

1. On the Division of Knowledge. Anticipation and Induction

OF human knowledge there are two main branches which may be called formal and material knowledge respectively. Their difference can be superficially described by saying that formal truth is tautologous and formal falsehood (self-)contradictory, whereas material truth consists in agreement and material falsehood in disagreement with facts. Formal truth is also called necessary, logical or analytic; material truth contingent, empirical or synthetic.

The possession of material knowledge may summarily be said to depend upon three basic faculties of our cognitive life, *viz.*, memory, observation, and anticipation. We remember facts which have been, observe (or experience) facts which are, and anticipate facts to come. Whether and to what extent these three faculties can be sharply separated and defined independently of one another, is a problem which we shall not discuss here.

Of anticipation we can distinguish two types, according to whether the anticipatory activity is dependent upon information afforded by observation and memory, or not. Anticipation of the first kind we shall call anticipation from experience, or induction. Whether and in what sense of the above word 'dependent' non-inductive anticipation actually occurs, is another problem which we pass over.

Of anticipation from experience there are again two sub-types.

The first sub-type may, as a first approximation, be described as follows: From the information that something has been the

13

case under certain conditions *and* that the conditions are repeated we anticipate that the same thing will be the case again. Anticipation of this kind which occurs in animals as well as in men, we shall call induction of the first order.

Induction of the first order can be studied from several points of view. Here we mention four:

First, it can be studied from an ethological or behaviouristic point of view, as the acquisition of habits or characteristic responses to stimuli. This study, being part of the psychology of learning, may, at least in the simplest cases, successfully employ the pattern of conditioned reflexes, roughly in the sense of Pavlov and Watson.

Secondly, it can be studied from a neurological point of view, as the acquisition of certain functional and possibly also structural units by the nervous system, which are causally responsible for the functional units of molar behaviour or habits.

Thirdly, it can (sometimes) be studied from a phenomenological or introspective point of view, as expectation, belief, and related states of consciousness, arising in the mind under certain conditions.

Fourthly, it can be studied from a logical point of view, as a sort of inference or argument. From the propositions that something has been the case under certain conditions *and* that the conditions are repeated we *infer*, as we say, the proposition that the same thing will be the case again. The conclusion of such an argument we shall call a prediction.

The second sub-type of anticipation from experience might, as a first approximation, be described as follows: From the information that something has been the case under certain conditions we anticipate that, *if* the same conditions are repeated, *then* the same thing will be the case again. Anticipation of this description, which seems to be characteristically human, we shall call induction of the second order.

Induction of the second order can also be studied under the behaviouristic aspect of habit, the neurological aspect of interconnexions in the nervous system, and the phenomenological aspect of expectation, though this study seems to be of a much more subtle and complicated nature than in the case of induction of the first order. Whether, from the points of view of psychology and nervous physiology, the two sub-types of antici‑ patory activity are fundamentally different, or whether the second cannot somehow be reduced to the first, is a debatable question.

Finally, induction of the second order can be studied under the logical aspect of inference. From the proposition that some‑ thing has been the case under certain conditions we *infer*, as we say, the proposition that, *if* the same conditions are repeated, *then* the same thing will be the case again. The conclusion of such an argument we shall call a theory.

Of theories we can distinguish two basic types, according to whether the inductive conclusion is intended unrestrictedly to apply to all cases in which the conditions are fulfilled, or to a restricted number of such cases only. Theories with numerically unrestricted range of application we shall call laws (Laws of Nature).

(It is clear that our use of the words 'theory' and 'law' cannot pretend to be co-extensive with all actual uses of them in ordinary language or scientific discourse.)

Given a theory and a proposition to the effect that, in a particular case, certain conditions are fulfilled, a prediction can be deduced. It may be regarded as one of the chief purposes of making theories, to provide thereby a basis for the deduction of predictions.

In this book the study of induction will be mainly confined to the logical aspect of inference.—It should be stressed that, in ordinary life as well as in science, conscious inference by no means always accompanies anticipatory activity. The logic, therefore, of making predictions and theories is not so much a study of actual intellectual procedures as of certain 'ration‑ alized counterparts' of them. This fact, however, does not diminish the importance of logical study for a true understand‑ ing of one of the principal aspects of human cognition.

2. On the Division of Science. The 'savoir pour prévoir'

Science, or the methodical search for and attainment of know‑ ledge, is sometimes divided into formal and material science, according to whether its aim lies in one or the other of the two main branches of human knowledge which we distinguished at the beginning of the previous section.

Material science may be further subdivided into theoretical and descriptive science. The distinguishing feature of theoretical science is the anticipation of facts from experience. Descriptive science is the systematic recording of facts.

Without claiming to be exhaustive, the above divisions exhibit certain types of scientific activity of which the actual sciences are a mixture. Any one of the actual sciences has, as a rule, a formal as well as a material, and a theoretical as well as a descriptive component. Purely formal, purely theoretical, and purely descriptive science thus represent borderline cases of a three-dimensional scale for the typological arrangement of actual sciences.

Formal science is sometimes called deductive. The anticipation of truth from experience is but another word for induction. Theoretical material science may therefore be called inductive.

The aim of science has frequently been characterized—by the words *savoir pour prévoir* or some equivalent phrase—as being to provide a basis for (successful) predictions. It is plain from the above that, strictly speaking, this applies only to that form of material science, which is here called theoretical. The characterization mentioned is not immediately applicable to formal science. Nor is it valid for descriptive material science, except in the normative sense of setting up an ideal. To what extent such a normative claim is justifiable will not be discussed here.

The above characterization of the aim of science is thus an exaggeration. The emphasis which has been laid on it is, however, illustrative of the important position which induction holds in the realm of scientific activity.

3. Induction and Discovery

Any act of anticipation from experience may be said to have been preceded by an act of discovery. From the failure to

distinguish sharply between anticipation and discovery mis-
understandings and disputes have arisen.

Discovery, as a preliminary phase of induction, consists in
the detection of a feature common to a set of particular data.
Anticipation, on the other hand, is the extension of this feature
to unexamined cases.

Sometimes the process of discovery is altogether trivial.
Such is, *e.g.*, the case when, from the fact that some ravens are
black, we anticipate the blackness of all ravens. When such
'primitive' instances of inductive inference are regarded from
the point of view of logic, one is apt wholly to ignore the
component of discovery, all the emphasis being laid upon the
aspect of anticipation.

In more 'advanced' cases the opposite tendency becomes
visible. Consider, *e.g.*, the well-known law for the refraction of
light $\frac{sin\ \alpha}{sin\ \beta} = c$. Before Snell, pairs of corresponding angles
α and β had been recorded in tables. Snell detected a feature
common to all such recorded correspondences. His law,
however, embodies much more than the discovery of this
feature; it also provides a rule which makes the prediction of
new pairs of corresponding angles possible. Here the anticipation follows as a trivial matter of course, the discovery having
once been made.

An example frequently mentioned in the literature on induction,[1] is Kepler's series of conjectures as to the orbit of Mars.
Observation had informed Kepler of the position of the planet
at various points in its path, and from this information the
orbit itself was to be 'induced'. After having tried and rejected,
so we are told, no less than nineteen assumptions as to the true
orbit, he finally discovered the law which agreed with the
observations. But, as Mill rightly pointed out in his polemic
against Whewell concerning the nature of induction,[2] the mere
discovery that the observed positions of Mars were correctly
represented by points on an imaginary ellipse, did not constitute
an inductive inference. It became an induction only when

[1] For a detailed analysis see Apelt, *Die Theorie der Induction* (1854), pp. 19–24
and 144–9.
[2] Mill, *A System of Logic* (1843), book. III, chap. II, §3.

combined with the assumption that Mars would continue to revolve in that same ellipse, and that the positions of the planet during the time which intervened between two observations must have coincided with the intermediate points of the curve. This, however, is obscured in saying, as we usually do, that Kepler discovered the form of Mars' orbit.[1] Laws of Nature, strictly speaking, are not discovered, but established through a process of inference on the basis of discoveries made.

The making of discoveries in science is a worthy object of systematic study. It can be pursued from several points of view.

An important aspect of discovery is the phenomenological, by which we shall understand the scientists' own records of how their discoveries have come into being as a result of mental activity and effort. Valuable contributions to the phenomenology of discovery have been made by great men of science like Helmholtz and Mach, Claude Bernard and Poincaré, Faraday and Darwin.

In addition to this aspect of discovery from within, so to speak, there is also the psychological and sociological aspect from without. What are the main features of the discoverer's intelligence? What are his habits of working? What influence upon his ingenuity is to be attributed to factors of heredity, social environment, education, etc.?[2]

Finally it may be asked, whether discovery can or cannot be studied from the point of view of logic. We need not attempt to answer the question here. We shall only draw attention to two traditional mistakes which have originated from a confusion of the logical aspect of anticipation with that of discovery.

The first mistake may be schematically described as follows:

The possibility of a Logic of Induction, studying the inferential mechanism of inductive reasoning, is thought to imply the possibility of a Logic of Discovery, providing rules or precepts for the invention of truths in science. This mistake has deep roots in the history of philosophy. It underlies the famous

---

[1] Apelt (*op. cit.*, p. 19) speaks of 'die logische Form jener merkwürdigen und klassischen Induction, durch welche Keppler die wahre Figur der Marsbahn entdeckte und welche stets als Muster und Vorbild des inductorischen Verfahrens in der Naturforschung gegolten hat.'

[2] Cf. the works of Jacques Picard, *Essai sur les conditions positives de l'invention dans les sciences* (1928) and *Essai sur la logique de l'invention dans les sciences* (1928).

Lullyan *Ars magna* which was also supposed to be an *ars inventiva veritatum*. It is, in particular, connected with the universal reaction in the sixteenth and seventeenth centuries against Aristotelian logic as being 'useless.' Descartes, Pascal, and the authors of the so-called Port Royal Logic are typical representatives of the view that logic ought to be somehow 'inventive.' In much the same spirit Zabarella, Galileo, and Leibniz expressed their opinions on induction and methodology. But by far the most important example of the confusion in question is found in the writings of Francis Bacon, who was strongly convinced that his contributions to the logical study of inductive inference were to make scientific discovery as independent of 'the acuteness and strength of wits' as had the invention of the compass made the drawing of exact circles independent of 'the steadiness and practice of the hand.'

One hundred years ago, William Whewell expressed in several works on induction the opinion that discoveries are, as he put it, 'happy guesses' or 'leaps which are out of the reach of method.' At the same time, Claude Bernard stressed similar views in polemics against Bacon.

As a criticism of the confusion of the logical study of scientific arguments with the invention of scientific truths, the attitude of Whewell and Claude Bernard towards discovery is fundamentally sound. If over-emphasized, however, it easily becomes unjustly discouraging to the systematic study of discovery as a psychological phenomenon. '*Il ne faut pas exagérer le mystère, ni faire de l'invention un miracle,*' as Picard[1] rightly says. And if carried to the extreme, it promotes a new confusion as regards the relation of induction to discovery. This second mistake may be schematically described as follows:

The impossibility of a Logic of Discovery, which ought to help us to find scientific truths, implies also the impossibility of a Logic of Induction, which gives an account of the logical mechanism of anticipation from experience. This mistake is obviously connected with the reaction in the last hundred years against unwarranted claims for logic in the realm of discovery. The reaction, as such, is undoubtedly sound and reflects a new and deepened understanding of the nature of the logical. It has,

Essai sur les conditions positives de l'invention dans les sciences, p. 7.

however, unduly neglected the inferential aspects of induction and must therefore to some degree be held responsible for the regrettable fact that, in spite of the enormous development of logic in recent times, the logical study of inductive procedures has not advanced very much from the state in which Mill left it.

Once the difference between anticipation and discovery is clearly apprehended, it should also be evident that past controversy over the possibility and nature of a Logic of Discovery has no direct bearings on the discipline which is the main theme of this book and for which 'Logic of Induction' is the appropriate name.

4. The Justification of Induction

Aristotle, who made the first attempt at a systematic treatment of induction, was already aware of the fact that inductive inference in our sense of the word has a non-demonstrative or inconclusive character. But Hume was the first to realize the full importance of this feature, in seeing that any type of argument, by means of which 'we can go beyond the evidence of our memory and senses,' is inductive and hence inconclusive. In view of this, it became a problem of first-rate importance, how to justify inductive inferences and, in particular, belief in the material truth of inductive conclusions. This is, in short, the problem of the Justification of Induction which is also rightly called the Problem of Hume.

The problem of how to provide a justification of induction has been the object of a vast and, in part, highly perplexing discussion in philosophy since the days of Hume. The 'sceptical' conclusions of Hume as to the possibility of such a justification have been difficult for philosophers to tolerate. The gigantic labours of Kant in the philosophy of knowledge were essentially devoted to showing that Hume was mistaken. Since Kant's efforts, on this crucial point at least, were not very successful, it has become the burden of succeeding philosophical systems to 'save' induction from Humean scepticism. Solutions were offered by Fries and the neo-Friesians, by Bradley and the English neo-Hegelians, by Meyerson, Whitehead, and many others. In contemporary philosophy new escapes have been proposed from the 'intellectual suicide'[1] recommended to science by Humeanism. The problem of justification transcends the realm of a logical study of induction as understood in this book. No sharp boundaries can, however, be drawn. But it is important to be aware of two traditional mistakes analogous to those mentioned in the previous paragraph as confusing the aspects of discovery and anticipation.

The first mistake is to think that the possibility of a Logic of Induction would imply a complete solution of the Problem of Hume. This mistake may be said to be implicit in Mill's writings on induction and in a vast number of text-books on the subject which follow his lines. It can be attributed to the joint influence of the failure clearly to apprehend that the canons of inductive logic cannot raise inductive arguments to conclusive power, and of the deep-rooted deductivistic ideal that legitimate reasoning must ultimately be demonstrative.

The second mistake arises when the alleged impossibility of solving the Problem of Hume is thought to imply the impossibility of a Logic of Induction. This mistake is typical of certain earlier writers of the logical empiricist trend of thought. It can be attributed to the joint influence of a clear insight into the non-demonstrative nature of induction, and of that same deductivistic ideal, just mentioned, which off-handedly identifies inconclusive reasoning with alogical argument.

* * * * *

The justification of Induction constitutes, it would seem, a 'philosophical' problem in a peculiar sense. What is meant by this can be imperfectly explained by saying that it is a problem, the essential issue of which is to make clear wherein the problem itself consists. Many of the perplexities with which it has been traditionally loaded, seem to arise from the fact that the disputants have entered the discussion without a clear idea as to the goal which was to be attained.

A suitable way of attacking the problem will therefore be to discuss the meaning of the word 'justification.' This discussion will reveal that there are, in fact, several different senses in

---

[1] Reichenbach, *Experience and Prediction* (1938), p. 344.

which a proposed justification of induction can be understood. Here we mention some.

i. The step which takes us from the premisses of an inductive inference to the conclusion is of a non-demonstrative nature. The converse step, however, is, in the case of induction of the second order, demonstrative: from the conclusion the premisses follow. This asymmetry is of some importance for the justification of theories and laws.

It would be an oversimplification of facts to think that knowledge of the premisses of an inductive inference is always prior in time to the introduction of the conclusion. The priority in question normally prevails, if the discovery of the feature which is common to a set of given data presents no difficulties. If, however, this is not so it may happen that we first, as the result of guesswork, hit upon the theory or the law and thereupon establish the premisses as the result of deduction from the suggested conclusion and subsequent verification of the deduced results through observation. The difficulty of not knowing the premisses until after reaching the conclusion is peculiar to situations in which the raw material, so to speak, for the construction of the inductive inference is afforded by data of measurement which have to be interpreted in terms of mathematical relationship (functionality). Good examples would again be afforded by Snell's law for the refraction of light and Kepler's laws for the movements of the planets. The recorded pairs of angles or the observed positions of Mars did not, as such, constitute premisses for an inductive inference. They first became premisses when it was shown that they possessed the feature prescribed by the respective laws in question.

If we thus succeed subsequently in showing that certain data afford premisses for a proposed conclusion, it is very natural indeed to say that our suggestion as to the conclusion has now been justified. This is an important sense in which we may and actually do talk of the justification of inductions in science.

We shall call this approach to the problem of justification the 'inventionistic' approach. It is intimately associated with a main trend in the history of scientific ideas which we may very roughly describe as the hypothetico-deductive. Among representatives of this trend of thought are found, beside professional philosophers, some also of the foremost innovators of theoretical science. One of its spiritual ancestors was Plato. His famous idea of 'saving the phenomena' was a sort of *carte blanche* justification for the making of theories and laws (according to a certain general pattern), from which the observed phenomena of nature could be deduced. It was partly against the misuse of this idea that Newton's *hypotheses non fingo* was directed.

In philosophy, the most emphatic spokesman of the inventionistic approach to the problem of justifying induction is, no doubt, William Whewell. In the opinion of Whewell, the deduction of the premisses of an inductive inference from the conclusion constitutes 'the criterion of inductive truth, in the same sense in which syllogistic demonstration is the criterion of necessary truth.'[1] Thus 'deduction justifies by calculation what induction has happily guessed.'[2] Induction and deduction are, in a characteristic sense, inverse operations of the mind.[3]

Against the background of a clear conception of the difference between discovery and anticipation it is not difficult to see how far the inventionistic approach will take us. The justification which it provides concerns the *discovery* of a common feature among given data, this discovery being first tentatively put forward in the shape of a proposed inductive conclusion. But it certainly does not give a justification for the act of *anticipating* the same feature in new data, which is also involved in putting forward the conclusion. For the mere fact that the given data of observation fall under the theory or law and thus, in this sense, justify its introduction, is no guarantee that the theory or law will hold also for future observation.

It is interesting to note that there is a tendency among supporters of this view of the philosophy of induction to compensate for the failure of the inventionistic approach to justify anticipation by taking up a conventionalist attitude as regards the universal truth of inductive conclusions. This was, *e.g.*,

[1] Whewell, *Novum Organon Renovatum* (3rd Ed. 1858), p. 115.
[2] *Ibid.*
[3] The view of induction and deduction being inverse operations has chiefly become associated with the name of Jevons. It should, however, be observed that Jevons did not confuse the verification of data deduced from the law with the verification of the law itself. This point has been overlooked by several critics of Jevons's opinions on induction (Erdmann, Meinong, Venn).

what Whewell did.[1] The conventionalistic attitude constitutes another approach to the problem of justification and will be discussed presently.

ii. It frequently occurs in advanced branches of theoretical science that data of observation suggest a certain uniformity which strikes us because we do not know how it fits in, so to speak, with other Laws of Nature. It may be that at first we even refuse to give it the name of a 'law' and refer to it merely as an 'empirical rule' or something of that sort. Later on it happens that we are able to 'explain' the rule, *i.e.*, we succeed in showing that it is, in fact, just a logical consequence of an already accepted law. This deductive connexion then serves as a justification of our first tentative inference from the observations to the uniformity.

As an example we may take Bradley's account of the aberration of the light of the fixed stars. Measurement of the yearly parallax of two stars led Bradley to the discovery of an (apparent) movement of the stars, the so-called ellipse of aberration. The rule for this movement was an inductive conclusion from the observations. Later Bradley succeeded in accounting for the movement as a joint effect of the movement of the earth and the finite velocity of light. This time his rule followed as a deductive conclusion from other Laws of Nature.

The above example illustrates another important sense in which we may and actually do, talk of the justification of inductions. We shall call this approach to the problem the deductivistic. Its most prominent spokesman in philosophy is perhaps E. F. Apelt. His little book *Die Theorie der Induction* (1854) still deserves attention; like the writings of William Whewell, it bears witness to its author's close contact with the realities of scientific practice. According to Apelt, laws are inductively inferred from particular observations and deductively justified from more general laws.[2]

---

[1] Cf. my thesis *The Logical Problem of Induction* (1941), pp. 70 and 215.

[2] Apelt, *op. cit.*, pp. 74–5: 'Die Induction bringt also nur die Untersätze des theoretischen Lehrgebäudes. In der vollendeten Theorie müssen diese Untersätze auf doppelte Weise festgestellt werden: einmal theoretisch oder, wie die Engländer sagen, deductiv d. i. als Lehrsätze, die durch systematische Ableitung aus ihrer Prinzipien folgen, das anderemal inductiv als Erfahrungssätze, die aus der Combination der Thatsachen folgen. Das erstere ist die Aufgabe der mathematischen Naturphilosophie, das andere die eigentliche Aufgabe der Naturforschung.'

The deductivistic approach may be said to provide a justification of the anticipatory aspect of law. It is, however, clear that this justification has a purely relative value. It makes the truth of some inductive conclusions depend upon the truth of others. The psychological importance of such a concatenation of our beliefs may nevertheless be considerable.

It is interesting to note that there is also among supporters of the deductivistic approach to the justification problem a tendency to compensate for its relativity by taking up a conventionalistic line as regards the question of the truth of the laws from which the deductions are made. This was clearly the case with Apelt.[1] It should, however, be observed that if the supreme laws are true by convention, then the inductively established laws which are subsequently deduced from them also become true by convention.

iii. The man of science, if asked what it is that justifies him in drawing a certain inductive conclusion from his observations, would probably answer in a great many cases by pointing neither to the fact that the observations can be deduced from the conclusion, nor to the fact that the conclusion is a consequence of other laws, but to the fact that the observations have been made in a certain methodical way so as to support the emergence of *this* conclusion, as opposed to other ones, from the data affording premisses to the inductive argument. This answer to the question of justification is particularly close at hand whenever the scientific problem is one of tracing a causal connexion in a bundle of phenomena.

This means that the justification of the inductive conclusion lies in the conformity of the argument to certain methodological standards or canons of induction. We shall call this approach to the problem of justification the inductivistic approach. The description of the standards involved constitutes one of the chief tasks of the Logic of Induction. (Cf. below Chap. IV.)

It has long been clear to philosophers that no inductive canons, however skilfully employed, can make the conclusion emerge with certainty from the premisses of the inductive argument, unless combined with some principles which themselves transcend memory and observation. These principles

---

[1] Cf. *The Logical Problem of Induction*, pp. 38–39.

25

are usually mentioned under names like 'the Law of the Uniformity of Nature' or 'the Law of Universal Causation.' It is obvious that unless we can establish these principles as true on some other basis than that of induction, then the relativity of the deductivistic approach to the problem of justification will attach to the inductivistic approach also.

There have been many efforts to establish the truth of some supreme principles justifying induction. Kant, broadly speaking, tried to show that these principles are embodied among the very preconditions of knowledge. Knowledge, as he puts it, would not be 'possible,' if some such principles were not true. Since, so he seems to reason, knowledge *is* possible, the principles *must be* true. Here, however, he has begged the question.[1] It can, moreover, be shown that even if Kant had been successful in trying to establish his preconditions of knowledge, he would not have succeeded in justifying induction, because the principles which he had in mind were not strong enough to enable us to raise any particular inductive conclusion to the rank of certainty.[2] The same is largely true of other attempts undertaken for a similar purpose. (Cf. below Chap. V.)

iv. We now mention the conventionalistic approach to the problem of justification. Its essence can be described as follows :

The leap from the premisses to the conclusion in inductive inference is a leap from the world of material into the world of formal knowledge. If it were not so, we could never be sure that future experience will not refute the conclusion. By taking the conclusion as true by convention, *i.e.*, by making the presence of the feature which is common to the data of observation a defining criterion of the presence of the conditions under which this feature will be repeated, it becomes irrefutable by experience.

It is obvious that such a transition from the material to the formal sphere of knowledge frequently takes place in connexion with inductive reasoning. We shall have occasion to return to this point later. (Cf. below Chap. VI.) It is, moreover, reasonable to assume that this transition, whether conscious or uncon-

---

[1] Cf. *The Logical Problem of Induction*, pp. 27–33.
[2] Cf. *ib.*, pp. 33-5.

scious, often actually serves as a justification of our belief in the *semper et ubique* of inductively established Laws of Nature.

The conventionalistic attitude towards the justification problem was something of a philosophic fashion at the beginning of this century. The first to point out the important rôle of convention in the formation of natural law was Henri Poincaré. Poincaré, however, was also clear about the limitations of this view of induction. Not so always the representatives of what may here be called radical conventionalism (Cornelius, Dingler, Schuppe, le Roy).

As already observed (cf. above p. 15), a theory or law can be used for predicting that, certain conditions being fulfilled, a certain feature will be observed. If such a prediction fails, then it must, from the conventionalistic point of view, be because either the fulfilment of the conditions prescribed by the law or the absence of the feature is only 'apparent' in the case in question. Thus conventionalism explains why unsuccessful predictions need not be interpreted as refutations of the theories or laws from which they were made. But conventionalism does not explain, why, in a great many cases, predictions actually *are* successful and why, as a rule, we do not resort to the conventionalistic remedy against refutation. Here the question of the justification of induction recurs and demands a new answer.[1]

v. Inductive inference, so long as it does not transcend the boundaries of material knowledge, cannot raise its conclusions to demonstrative certainty. But it is thought capable of conferring upon them a lower or higher degree of probability. This fact is supposed to cast new light on the problem of justification. —Inductive inference is frequently also called probable inference and contrasted with deductive inference which is called certain.

Of the probabilistic approach to the justification problem there are two variants which must be carefully kept apart.

The first variant assumes that the justification of induction in terms of probability should provide some sort of guarantee that inductive predictions will be successful, if not immediately, at least 'in the long run.' A typical representative of this view is Reichenbach.

[1] Cf. *The Logical Problem of Induction*, pp. 48–64 and 209–13.

There is a straightforward way of making probability imply predictive success, *viz.* by accepting the so-called Frequency Interpretation of the concept of probability. On this interpretation, however, propositions about probabilities themselves become inductive or anticipatory.

The second variant refuses to accept success in predictions as a standard by which to justify induction. 'The validity of the inductive method does *not* depend on the success of its predictions,' says Keynes,[1] the most conspicuous advocate of this opinion.

Of the positive characteristics of this second variant it will suffice here to mention that it takes probability either as a notion which cannot be made explicit in terms of other ideas, or as interpretable in terms, *e.g.*, of possibility or belief, which do not necessarily make propositions about probabilities inductive.

It appears that both variants can be worked out in a manner which gives them undeniable merits as contributions to the justification problem. The following point, however, should be carefully observed:

The first variant, if it accepts the Frequency Interpretation, can ultimately justify induction only in a relative sense which makes it possible to raise the same problem on a new level. The second variant again, whether or not it accepts an interpretation of probability, proves nothing about the future course of things. Thus, on the interpretation mentioned, the first variant shares a characteristic weakness with the deductivistic and the inductivistic theories, and the second a similar weakness with the inventionistic and conventionalistic approaches to the problem of justification.

The crucial question has been whether a solution to the Problem of Hume could be found, which would assure us of success in prediction without begging the question. The traditional view of the eighteenth and nineteenth centuries was that there did in fact exist such a solution, in virtue of certain theorems of the Calculus of Probability known as the Principles (Laws) of Great Numbers. These laws seemed somehow to make possible the deduction of frequencies 'in the long run' from non-inductive assumptions about possibilities. The

---

[1] *A Treatise on Probability* (1921), p. 221.

fallacy of this reasoning was for the first time exposed with full clarity by Robert Leslie Ellis in his important paper *On the Foundations of the Theory of Probabilities* (1842). To-day the essentials of Ellis's criticism are universally accepted.[1]

Generalizing this criticism we may say that if a guarantee of success in predictions is to be established by deduction, then it must rely on premisses which themselves involve inductive assumptions. On the other hand, it is difficult to see what other way, besides the deductive, of foretelling success would really deserve the name of a 'guarantee.'[2] Hence it would appear that the idea of merging the two variants of the probabilistic approach to the justification problem into one doctrine combining the advantages of both, is an illusion.

* * * * *

We have now seen that there are several different ways in which we may and actually do talk of a justification of induction. Each way answers to a certain practical demand which may be felt in connexion with inductive reasoning. None of them, it seems, answers to all such demands. In the concrete situation, as it occurs in science or in practical life, the various justifications are valued for their merits of meeting one definite demand rather than criticized for their demerits of not meeting others. The scientist who has hunted for the cause of some interesting phenomenon and claims to have found it, will certainly, in the eyes of himself and his colleagues, have justified his claims, if his experiments and observations are shown to have been up to the current standards of accuracy and methodical performance. The validity of universal causation and other assumptions on which he must rely if his conclusion is to be indisputably true, do not even enter the question. Nor would he necessarily regard his conclusion as unjustified even if it later turned out to be false.

It is only when the problem of justification is raised so to say 'in vacuo,' *i.e.*, not in relation to a concrete situation, that the demerits of the respective justifications first become obvious and the philosophic problem discussed with so much fervour in the last 200 years becomes urgent. This detachment from

---

[1] Cf. *The Logical Problem of Induction*, pp. 159–64 and 237–40.
[2] Cf. *ib.*, pp. 155–9, 164–9, 236–7, and 240–2.

practical thinking is, of course, no sign that the problem is un-important. But it is a point which deserves some attention.

The analysis of the word 'justification' and the working out of alternative senses in which induction may be said to be or not to be justifiable, is a procedure which will ultimately take us to the very foundations of human knowledge, to the discussion of expressions like 'to know,' 'to believe,' 'to be certain,' 'to prove,' *etc.* If a solution of the Problem of Hume is expected once and for all to bring out the whole relevance of this dis-cussion to induction, then claims as to a 'solution' already seem in principle inappropriate and narrow-minded.

But the philosophic situation can also be regarded from another point of view. As may be seen from the sketchy remarks earlier in this section, the working out of the various justifica-tions of induction shows the demerits as well as the merits of the alternatives. Each alternative permits certain conclusions in regard to justification and does not permit certain others. The alternatives, moreover, seem to be of two basic types, depending upon whether they permit demonstrative conclusions concern-ing success in predictions, or not. It is characteristic of the former alternatives that they are based on premises which are themselves inductive, whereas the latter need not go beyond the limits of actual observation. It is not unplausible to assume that some of the intellectual discomfort traditionally associated with the Problem of Hume has been nourished by attempts to establish a guarantee of predictive success on non-inductive premises without a full awareness of the fact that a 'guarantee' cannot then mean the same as a demonstration. In so far as clarity on this important point can make the discomfort vanish, it would not be inappropriate to talk of a solution to the Problem of Hume attained through an analysis of the meaning of 'justification.'

5. *The Three Problems of Induction*

We have in this chapter tried to distinguish between three main problems of induction which have been traditionally intertwined: The mainly psychological problem of discovery or of the origin of inductive inferences in science, the logical problem of analysing the inferential mechanism of induction, and the specifically philosophical problem of justification.

The problems of discovery and justification will not be further discussed in this book. With the Justification of Induction I have dealt at length in a previous publication, to some passages of which I have ventured to refer above.[1]

The main topic of this book is the logic of inductive inference. The logic of Induction may conveniently be divided into two parts, depending on whether the question of the truth and falsehood of the conclusions alone enters the discussion, or whether there is the further question of the probability of the conclusions also.

The first part may be called the Logic of Inductive Truth. It is treated in Chaps. IV, V, and VI.

The second part may be called the Logic of Inductive Probability. It is treated in Chaps. VIII, IX, and X.

The apparatus and technical vocabulary of logic needed for the investigation are presented in an introductory chapter (Chap. II.) This apparatus is used in Chap. III to restate in more precise terms the logical form of inductive arguments and their various parts, as loosely outlined at the beginning of the present chapter.

A special chapter (Chap. VII) is devoted to a general treatment of probability.

* * * * *

We need not here dwell upon the old controversy over the 'possibility' of a Logic of Induction. This controversy has been largely due to the failure to separate the task of logical analysis from the tasks of promoting scientific discovery and of justifying our belief in induction respectively. The principal confusions on this point have been outlined above. Once the distinction has become clear, it will no longer be necessary to present an argument in defence of the possibility of studying induction from the point of view of logic. Investigation into the respective problems of induction can therefore be pursued unhampered by false pretensions and misleading expectations.

---

[1] The term 'logical problem of induction' was used in that book to cover the two problems which we have here distinguished as the problem of logical analysis and the problem of justification respectively.

Chapter Two

PRELIMINARY CONSIDERATIONS ON LOGIC

1. On Propositions

A PROPOSITION is that so-and-so (is the case). Propositions are expressed or symbolized by sentences.

The Logic of Propositions studies propositions as 'unanalysed wholes' and sentences as symbols of such unanalysed propositions.

Propositions can be related to truth-values. Truth-values are truth and falsehood.

For the relation of propositions to truth-values we have the following two principles:

  i. Every proposition is true or false.
  ii. No proposition is true and false.

A proposition is called a truth-function of n propositions, if there is a rule for the determination of the truth-value of the proposition for each combination of truth-values in the n propositions.

The following seven truth-functions are defined separately:

By the negation(-proposition) of a given proposition we understand that proposition which is true if, and only if, the given proposition is false. If a expresses a proposition, then a expresses its negation.

By the conjunction(-proposition) of two propositions we understand that proposition which is true if, and only if, both the propositions are true. If a and b express propositions, then $a \& b$ expresses their conjunction.

By the disjunction(-proposition) of two propositions we understand that proposition which is true if, and only if, at least one of the propositions is true. If a and b express propositions, then avb expresses their disjunction.

By the implication(-proposition) of a first proposition, called the antecedent(-proposition), and a second proposition, called the consequent(-proposition), we understand that proposition which is true if, and only if, it is not the case that the antecedent is true and the consequent false. If a expresses the antecedent and b the consequent, then $a{\rightarrow}b$ expresses their implication.

(Henceforth, in talking about implications, explicit reference to the order of propositions will not be made. The order is assumed to be clear to the reader from the context.)

By the equivalence(-proposition) of two propositions we understand that proposition which is true if, and only if, both the propositions are true or both are false. If a and b express propositions, then $a{\leftrightarrow}b$ expresses their equivalence.

By the tautology(-proposition) of n propositions we understand that proposition which is true for every combination of truth-values in the n propositions.

By the contradiction(-proposition) of n propositions we understand that proposition which is false for every combination of truth-values in the n propositions.

In virtue of i and ii above, every proposition is a truth-function of itself.

Truth-functionship is transitive. If a proposition is a truth-function of n propositions and every one of these is a truth-function of $m_1, \ldots, m_n$ further propositions, then the first proposition is a truth-function of the latter propositions.

It follows from the definition of a truth-function that if a proposition is a truth-function of n propositions, then it is also a truth-function of any $m+n$ propositions which include among themselves the original n propositions.

We call $\bar{a}$ the negation-sentence of a, and $a\&b$ the conjunction-, avb the disjunction-, $a{\rightarrow}b$ the implication, and $a{\leftrightarrow}b$ the equivalence-sentence of a and b.

A sentence which is neither the negation-sentence of another sentence, nor the conjunction-, disjunction-, implication-, or equivalence-sentence of two other sentences, is called an atomic sentence.

By a molecular complex of n sentences we understand:

i. Any one of the n sentences themselves and their negation-sentences.

ii. The conjunction-, disjunction-, implication-, or equivalence-sentence of any two of the n sentences.

iii. The negation-sentence of any molecular complex of the n sentences, and the conjunction-, disjunction-, implication-, or equivalence-sentence of any two molecular complexes of the n sentences.

The n sentences are called the constituents of their molecular complexes. If the sentences are atomic, they are called atomic constituents.

(It is sometimes convenient to call the propositions expressed by the n sentences, constituents of the propositions expressed by their molecular complexes.)

As to the use of brackets we adopt the conventions that the symbol & has a stronger binding force than v, $\rightarrow$, and $\leftrightarrow$; the symbol v than $\rightarrow$ and $\leftrightarrow$; and the symbol $\rightarrow$ than $\leftrightarrow$. Thus, e.g., if a, b, c, d, and e are sentences, we can instead of $(((a\&b)vc)\rightarrow d)\leftrightarrow e$ write simply $a\&bvc\rightarrow d\leftrightarrow e$.

In virtue of the transitivity of truth-functionship, any molecular complex of n sentences expresses a truth-function of the propositions expressed by the n sentences themselves. Which truth-function of the propositions expressed by its constituents a molecular complex expresses, can be investigated and decided by means of truth-tables. The technique of constructing truth-tables is supposed to be familiar to the reader.

A molecular complex which expresses the tautology (contradiction) of the propositions expressed by its atomic constituents will be called tautologous (contradictory) and said to express formal truth (falsehood) in the Logic of Propositions.

If two sentences express the same proposition, they are called identical. It is sometimes convenient to say that identical sentences express identical propositions. If a and b are sentences,

the proposition that they are identical may be expressed $a = b$.[1]
If two molecular complexes are identical, their equivalence-sentence expresses the tautology of the propositions expressed by their atomic constituents.

If one sentence is identical with the conjunction-sentence of itself and another sentence, the first sentence is said to entail the second. The second sentence is said to follow from the first. The relation of entailment is also said to subsist between the expressed propositions.

If the first of two molecular complexes entails the second, then their implication-sentence expresses the tautology of the propositions expressed by their atomic constituents.

If a and b and c are sentences, the following pairs of molecular complexes are identical:

i. a and $\bar{\bar{a}}$. (Law of Double Negation.)

ii. $a{\rightarrow}b$ and $\bar{b}{\rightarrow}\bar{a}$. (Law of Contraposition.)

iii. $a\&b$ and $\overline{\bar{a}v\bar{b}}$, and avb and $\overline{\bar{a}\&\bar{b}}$ respectively. (Laws of de Morgan.)

iv. $a\&b$ and $b\&a$, and avb and bva respectively. (Laws of Commutation.)

v. $(a\&b)\&c$ and $a\&(b\&c)$, and $(avb)vc$ and $av(bvc)$ respectively. (Laws of Association.)

vi. $a\&(bvc)$ and $a\&bva\&c$, and $avb\&c$ and $(avb)\&(avc)$ respectively. (Laws of Distribution.)

In virtue of the Laws of Association we can omit brackets from conjunctions of conjunction-sentences and from disjunctions of disjunction-sentences. Thus, e.g., we write $a\&b\&c$ for $(a\&b)\&c$ and $avbvc$ for $(avb)vc$. We can henceforth talk of the conjunction and disjunction of n propositions and of n-termed conjunction- and disjunction-sentences respectively.

Any molecular complex can be transformed into certain so-called normal forms.

---

[1] It should be observed that, whereas $a{\leftrightarrow}b$ expresses a 'relation' (equivalence) between the *propositions* expressed by the sentences a and b, $a = b$ expresses a 'relation' (identity) between the *sentences* a and b themselves. It might be suggested that we should write 'a' = 'b' and not $a = b$. We shall, however, avoid the use of quotes throughout.

Let there be n atomic sentences. We select m ($O \leqslant m \leqslant n$) of those sentences and the negation-sentences of the remaining n-m sentences and form their conjunction-sentence. The formation of the conjunction-sentence can take place in $n!$ different ways, but the conjunction-sentences thus formed are all identical, for which reason we regard it as immaterial which way is chosen. The selection of m sentences can again take place in $\sum\limits_{m=O}^{n} \binom{n}{m}$ or 2^n different ways. Thus we get 2^n conjunction-sentences, no two of which are identical.[1]

In a similar manner we form of those n atomic sentences 2^n disjunction-sentences.

If the two extreme cases of 0- and 1-termed conjunction- and disjunction-sentences are included, it can be shown that any molecular complex of the n atomic sentences is identical with a 0-, 1- or . . .or 2^n-termed disjunction of the above conjunction-sentences and also with a 0-, 1- or . . . or 2^n-termed conjunction of the above disjunction-sentences. The disjunction of conjunction-sentences we call its perfect disjunctive normal form in terms of the n atomic sentences, and the conjunction of disjunction-sentences its perfect conjunctive normal form.

If the molecular complex expresses the contradiction of the propositions expressed by its atomic constituents, its perfect disjunctive normal form vanishes, *i.e.*, is a 0-termed disjunction, whereas its perfect conjunctive normal form is made up of all the above 2^n disjunction-sentences. If the molecular complex expresses the tautology of the propositions expressed by its atomic constituents, its perfect conjunctive normal form vanishes, and its perfect disjunctive normal form is 2^n-termed.

The technique of finding the perfect normal forms of a given molecular complex will not be described here.

A proposition is said to be logically totally independent of n propositions, if, for no combination of truth-values in the n propositions, is there a rule determining the truth-value of the proposition.

n propositions are said to be logically totally independent,

---

[1] $n!$ means the product of the n first cardinals and $\binom{n}{m}$ is an abbreviation for $n! : m! (n-m)!$

if every one of the propositions is logically totally independent of the remaining n-1 propositions.

If a proposition is logically independent of another proposition, they are mutually logically independent. Be it observed, however, that any two of n propositions may be mutually logically independent without the n propositions being logically totally independent. The converse is not possible.

*   *   *   *   *

A sentence which does not express formal truth or falsehood, is said to express a material proposition.

We call a material proposition verifiable, if it is possible to come to know the truth of it, and falsifiable, if it is possible to come to know its falsehood.

According to a certain opinion among philosophers, no material proposition is ultimately verifiable or falsifiable. The proposition is accepted or rejected on the basis of certain evidence in its favour or disfavour, but this evidence is never thought sufficient for complete verification or falsification respectively. There is always the possibility that the accumulation of further evidence will affect a change in our attitude as regards the truth-value of the proposition.—It should be observed that the discussion throughout this book is neutral in respect of this opinion.

2. On Properties

Propositions can be analysed into parts which are not themselves propositions. There are two principal ways of analysis. The first way we call the Aristotelian view of propositions. According to it, to assert a proposition is to attribute a property to a thing (an object, an individual). The 'thing' may itself be a property. The second we call the relational view of propositions. According to it, to assert a proposition is to assert a relation between a number of things.

We shall here assume that every proposition *can* be analysed in the Aristotelian way.

Let there be a proposition which is analysed in the Aristotelian way. If the proposition is true, we say that the property is

37

present in the thing and call the thing a positive instance of the property. If the proposition is false, we say that the property is absent from the thing and call the thing a negative instance of the property.—Presence and absence (of a property in a thing) are called presence-values (of the property in the thing).

It will be convenient to adopt some rudimentary form of a Theory of Logical Types. The reasons and the justification for this will not be discussed here.

The positive instances of a property are said to constitute the extension of the property.

The extension of a property is a set or a class. The positive instances of the property are also called members of the set. Whether every set or class is also the extension of a property need not concern us here. It should, however, be observed that whenever we introduce a name for a specified set, this name is treated as the name of a property, *viz.*, the property of being a member of the set in question.

The (positive and negative) instances of a property are said to constitute a Universe of Things. The properties which are present or absent in a thing are said to constitute a Universe of Properties.

It will be assumed that each member of a given Universe of Things constitutes the same Universe of Properties, and *vice versa*. This enables us to speak of corresponding universes (of things and properties).

Henceforth in talking about properties and things in the same context we always tacitly assume that they are properties and things from corresponding universes.

On the above assumptions as to division into universes, we obtain the following two principles for the relation of properties to presence-values from the corresponding basic principles of the preceding paragraph for the relation of propositions to truth-values:

i. Every property is present or absent in a given thing.
ii. No property is present and absent in a given thing.

A property is called a presence-function of n properties, if there is a rule for the determination of the presence-value of the property in a thing for each combination of presence-values of the n properties in the same thing.

The following seven presence-functions are defined separately:

By the remainder or negation(-property) of a given property we understand that property which is present in a thing, if and only if the given property is absent in the same thing. If A denotes, *i.e.*, is the name of, a property, then $\bar{A}$ denotes its remainder or negation.

Similarly we define the product or conjunction-, the sum or disjunction-, the implication-, and the equivalence-property of two properties. If A and B denote properties, then $A\&B$ denotes their product or conjunction, $A\mathrm{v}B$ their sum or disjunction, $A{\rightarrow}B$ their implication, and $A{\leftrightarrow}B$ their equivalence.

Finally, we define the tautology(-property) of n properties as that property which is present, and the contradiction(-property) of n properties as that which is absent, in a thing for every combination of presence-values of the n properties in the thing.

If properties are represented as squares within a square, we can represent any presence-function of them as a shadowed part of the square:

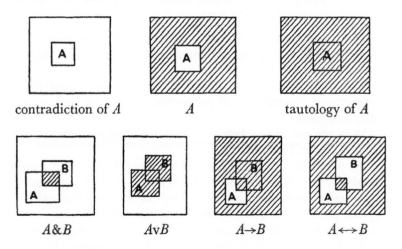

contradiction of A A tautology of A

$A\&B$ $A\mathrm{v}B$ $A{\rightarrow}B$ $A{\leftrightarrow}B$

In virtue of i. and ii. above, every property is a presence-function of itself.

Presence-functionship is transitive. (Cf. above p. 33.)

It follows from the definition of a presence-function that if a property is a presence-function of n properties, then it is also a presence-function of any $m+n$ properties which include among themselves the original n properties.

We call $\bar{A}$ the negation-name of A, and $A\&B$ the product- or conjunction-, $A\lor B$ the sum- or disjunction-, $A\to B$ the implication-, and $A\leftrightarrow B$ the equivalence-names of A and B.

A name of a property which is neither the negation-name of another name of a property, nor the conjunction-, disjunction-, implication-, or equivalence-name of two other names of properties is called an atomic name.

By a molecular complex of n names of properties we understand:

i. Any one of the n names themselves and their negation-names.

ii. The conjunction-, disjunction-, implication-, or equivalence-name of any two of the n names.

iii. The negation-name of any molecular complex of the n names, and the conjunction-, disjunction-, implication-, or equivalence-name of any two molecular complexes of the n names.

The n names are called the constituents of their molecular complexes. If the names are atomic, they are called atomic constituents.

(It is sometimes convenient to call the properties denoted by the n names, constituents of properties denoted by their molecular complexes.)

As to the use of brackets we adopt the same conventions as in the Logic of Propositions.

In virtue of the transitivity of presence-functionship, any molecular complex of n names of properties denotes a presence-function of the properties denoted by the n names themselves. Which presence-function of the properties denoted by its constituents a molecular complex denotes, can be investigated and decided by means of presence-tables. The technique of constructing presence-tables is strictly analogous to that of constructing truth-tables.

A molecular complex which denotes the tautology (contradiction) of the properties denoted by its atomic constituents will be called tautologous (contradictory) and said to denote formal presence (absence) in the Logic of Properties.

If two names denote the same property, they are called identical. It is sometimes convenient to say that identical names denote identical properties, and that names which are not identical denote different properties. If A and B are names of properties, the proposition that they are identical may be expressed $A = B$.

If two molecular complexes of names are identical, their equivalence-name denotes the tautology of the properties denoted by their atomic constituents.

If one name of a property is identical with the conjunction-name of itself and another name of a property, the first name is said to entail the second. The relation of entailment is also said to subsist between the denoted properties.

If the first of two molecular complexes of names entails the second, then their implication-name denotes the tautology of the properties denoted by their atomic constituents.

Analogous laws to those of Double Negation, Contraposition, de Morgan, Commutation, Association, and Distribution in the Logic of Propositions are valid in the Logic of Properties.

In virtue of the Laws of Association we can henceforth talk of the product and sum of n properties and of n-termed conjunction- and disjunction-names of properties respectively.

If there are n properties denoted $A_1, \ldots, A_n$, we shall use ΠA_n to denote their product and ΣA_n to denote their sum.

Further, given any set of properties (from the same universe), we can talk of the product of *all* members of the set, meaning the property of which it is true that a thing is a positive instance of it if, and only if, the thing is a positive instance of every property from the set.

Similarly, given any set of properties (from the same universe), we can talk of the sum of *all* members of the set, meaning the property of which it is true that a thing is a positive instance of it if, and only if, the thing is a positive instance of some property in the set.

The normal forms of molecular complexes of names of

properties are strictly analogous to the normal forms of molecular complexes of sentences. If the two extreme cases of 0- and 1-termed conjunction- and disjunction-names are included, any molecular complex of n names of properties has a perfect disjunctive and a perfect conjunctive normal form in terms of the n names.

The perfect disjunctive (conjunctive) normal form of a molecular complex in terms of n names might be called the perfect disjunctive (conjunctive) normal denotation, in terms of the n names, of the property denoted by the molecular complex.

Suppose that a property which is a presence-function of n given properties is not also a presence-function of m $(m<n)$ of those n properties. If this is true, then the property's perfect disjunctive (conjunctive) normal denotation in terms of atomic names of the n properties will be called a smallest perfect disjunctive (conjunctive) normal denotation.

Be it observed that the perfect disjunctive and the perfect conjunctive normal forms of a molecular complex in terms of n names are 'complementary' in the sense that if the first is an m-termed disjunction-name, the second is a 2^n-m-termed conjunction-name, and *vice versa*.

Consider n atomic names of properties and the 2^n names which we get by taking m $(0 \leqslant m \leqslant n)$ of those names and the negation-names of the remaining $n-m$ names and forming (order being irrelevant) their conjunction-name. (Cf. above p. 36.) If there are 2^n things which are positive instances of the respective properties denoted by the 2^n conjunction-names, we say that those things constitute the Total Variation in the realm of the properties denoted by the n atomic names.

A property is called logically independent of another property, if, given the presence-value of the second property in a thing, there is no rule determining the presence-value of the former in the same thing.

A property is called logically totally independent of n properties, if, for no combination of presence-values in the n properties, is there a rule determining the presence-value of the first property.

A property is called logically totally independent of (the

properties in) a set of properties (from the same universe), if it is logically totally independent of any n properties in the set.

n properties are called logically totally independent, if every one of the properties is logically totally independent of the n-1 other properties.

The properties in a set of properties (from the same universe) are called logically totally independent, if any n properties from the set are logically totally independent.

*   *   *   *   *

If no instance of a property is positive, the property is called empty.

If at least one instance of a property is positive, the property is said to exist.

If all instances of a property are positive, the property is called universal.

The proposition that the property denoted by A exists, will be expressed by the sentence $E\ A$.

The proposition that the property denoted by A is empty, is the negation-proposition of the above. It can thus be expressed by the sentence $\overline{E\ A}$. As an abbreviation for this sentence we shall use $\bar{E}\ A$.

The proposition that the property denoted by A is universal, we shall understand as the proposition that the negation of the property denoted by A is empty. It can thus be expressed by the sentence $\bar{E}\ \bar{A}$. As an abbreviation for this sentence we shall use $U\ A$.

Propositions to the effect that a property is empty, exists, or is universal we call quantified. The sentences expressing them are also called quantified. The symbols E and U are called quantifiers.

A proposition to the effect that a property exists is called an Existential Proposition. A proposition to the effect that a property is universal is called a Universal Proposition. According to the above, the negation of an Existential Proposition is a Universal Proposition, and *vice versa*.

If the implication-property of two properties is universal, the first property is said to be included in the second. If A and B

are names of properties, inclusion can thus be expressed by the sentence U $A{\rightarrow}B$. As an abbreviation of this sentence we shall use $A \subset B$.

If the equivalence-property of two properties is universal, the properties are called co-extensive. If A and B are names of properties, co-extension can thus be expressed by the sentence U $A{\leftrightarrow}B$. As an abbreviation of this sentence we shall use $A \equiv B$.

A proposition to the effect that a property is included in another property we call a Universal Implication, and a proposition to the effect that two properties are co-extensive a Universal Equivalence.

(Universal Implications and Equivalences are Universal Propositions.)

It is important not to confuse inclusion with implication and entailment, nor co-extension with equivalence and identity.

By an instance of an Existential Proposition we understand a proposition to the effect that a thing is a positive instance of the property the existence of which is being asserted (in any sentence expressing the Existential Proposition). By an instance of a Universal Proposition we understand a proposition to the effect that a thing is a positive instance of the property the universality of which is being asserted (in any sentence expressing the Universal Proposition).

The thing is said to afford the instance.

If an instance of an Existential or Universal Proposition is a true proposition, we call it a confirming instance. If it is a false proposition, we call it a disconfirming instance.

If the instance is confirming and if it entails the Existential or Universal Proposition, we call it a verifying instance. If the instance is disconfirming and if its negation entails the negation of the Existential or Universal Proposition, we call it a falsifying instance.

Any confirming instance of an Existential Proposition is a verifying instance, and any disconfirming instance of a Universal Proposition is a falsifying instance.

Unless the property, the existence or universality of which is being asserted, is known to have one instance only, no disconfirming instance of an Existential Proposition is a falsifying instance, and no confirming instance of a Universal Proposition is a verifying instance.

In order to avoid confusion, sentences using the symbols E, U, $\subset$, or $\equiv$ will be enclosed within brackets, if they occur as constituents of molecular complexes of sentences.

We lay down the following Principle of Existence:

If a property is the disjunction of n properties, then the proposition that the property exists is the disjunction of the proposition that the first of the n properties exists, and . . ., the proposition that the last of the n properties exists.

Let there be a molecular complex of sentences, the constituents of which express Quantified Propositions. Let the number of constituents be k and let the number of atomic names of properties which occur in the complex be n.

Consider the 2^n names which we get by taking m ($0 \leqslant m \leqslant n$) of those atomic names and the negation-names of the remaining $n\text{-}m$ atomic names and forming (order being irrelevant) their conjunction-name. Consider the 2^n properties denoted by these conjunction-names. The 2^n sentences expressing that the first of the 2^n properties exists, and . . ., and that the last of the 2^n properties exists, we shall call the existence-constituents of the molecular complex of sentences.

As observed above, any Universal Proposition is the negation of an Existential Proposition. Hence the molecular complex is identical with another molecular complex of k sentences, all of which express Existential Propositions.

Consider the k properties, the existence of which is expressed in the constituents of this second molecular complex of sentences. Any one of them has a perfect disjunctive normal denotation in terms of the above n atomic names. This means that any one of the k properties is a disjunction of 0, 1, or . . ., or all, of the above 2^n properties denoted by conjunction-names derived from the atomic names. Hence, in virtue of the Principle of Existence, any one of the k constituents of the second molecular complex expresses the disjunction of 0, 1, or . . ., or all, of the propositions expressed by the above 2^n sentences which we called existence-constituents.

Since truth-functionship is transitive, the second molecular complex of sentences expresses a truth-function of the propositions expressed by the 2^n existence-constituents. Since the second molecular complex is identical with the first, the first also expresses a truth-function of the propositions expressed by the 2^n existence-constituents.

Thus any molecular complex of sentences, the constituents of which express Quantified Propositions, expresses a truth-function of the propositions expressed by the existence-constituents of the complex. Which truth-function the molecular complex of sentences expresses can be investigated and decided by means of truth-tables.

A molecular complex of quantified sentences which expresses the tautology (contradiction) of the propositions expressed by its existence-constituents will be called tautologous (contradictory) and said to express formal truth (falsehood) in the Logic of Properties.

If two molecular complexes of quantified sentences are identical, their equivalence-sentence expresses the tautology of the propositions expressed by its existence-constituents.

If the first of two molecular complexes of quantified sentences entails the second, then their implication-sentence expresses the tautology of the propositions expressed by its existence-constituents.

3. On Relations

If x_1 and . . . and x_n are names of things and R is the name of a relation, then $R(x_1, . . ., x_n)$ will be used to express the proposition that the relation denoted by R subsists between the things denoted by x_1 and . . . and x_n.

If the number of related things is n, the relation is called n-adic or n-termed.

It is essential to the relational view of propositions that the things should be taken in a certain order.

An ordered set of n things will be called simply an order.

Let there be a proposition which is analysed in the relational way. If the proposition is true, we say that the relation subsists in the order, and call the order a positive order of the relation. If the proposition is false, we say that the relation does not subsist in the order, and call the order a negative order of the relation.—Subsistence and non-subsistence (of a relation in an order) are called subsistence-values (of the relation in the order).

The rudimentary form of a Theory of Logical Types adopted for properties will be extended to relations.

The positive orders of a relation are said to constitute its extension. The (positive and negative) orders of a relation are said to constitute a Universe of Orders.

The relations which subsist or do not subsist in an order are said to constitute a Universe of Relations.

The first members of each positive order of a relation are said to constitute its first positive domain, and the first members of negative orders only its first negative domain. The first members of each (positive or negative) order of a relation are said to constitute the first domain of the relation.

Similarly we define the second positive domain, negative domain, and domain of a relation. *Etc.*

It is assumed that the first, second, and further domains of a relation are Universes of Things in the sense defined in the Logic of Properties. (Cf. above p. 38.)

It is further assumed that the relations which subsist in the several orders of a given relation constitute the same Universe of Relations, and *vice versa* that the orders of each relation which subsists in a given order constitute the same Universe of Orders. This enables us to speak of corresponding Universes of Orders and Relations in the Logic of Relations.

Henceforth in talking about relations and orders in the same context we always tacitly assume that they are relations and orders from corresponding universes.

If the first, second and further domains of a relation constitute the same Universe of Things, the relation is called homogeneous, otherwise non-homogeneous.

On the above basis we obtain the following two principles from our initial principles for the relation of propositions to truth-values:

    i.   Every relation subsists or does not subsist in a given order.

    ii. No relation subsists and does not subsist in a given order.

A relation is called a subsistence-function of n relations, if there is a rule for the determination of the subsistence-value of the relation in an order for each combination of subsistence-values of the n relations in the same order.

In strict analogy to the corresponding truth- and presence-functions we can define the negation of a given relation, the product or conjunction, the sum or disjunction, the implication, and the equivalence of two relations, and the tautology and the contradiction of n relations.

Analogous symbols will be used.

We thereby define what is meant by an atomic name of a relation, and by a molecular complex of n names of relations.

Any molecular complex of n names of relations denotes a subsistence-function of the relations denoted by the n names themselves. Which subsistence-function it denotes can be investigated and decided by means of subsistence-tables.

A molecular complex which denotes the tautology (contradiction) of the relations denoted by its atomic constituents will be called tautologous (contradictory) and said to denote formal subsistence (non-subsistence) in the Logic of Relations.

Analogous laws to those of Double Negation, Contraposition, de Morgan, Commutation, Association, and Distribution in the Logic of Propositions and Properties are valid in the Logic of Relations.

Molecular complexes of names of relations have normal forms analogous to the normal forms of sentences and names of properties.

The idea of independence of relations is analogous to the idea of independence of propositions and properties.

The Aristotelian view of propositions can be applied to relational propositions in virtue of the following device:

An order can be treated as a thing and a relation which subsists in the order can be viewed as a property of the order.

This extension of the thing-property view to relational propositions is of some importance. It implies that whenever we talk of properties in the subsequent treatment of inductive inference and of natural law, the word 'property' should be understood to cover both properties in the genuine sense and relations viewed as properties of ordered sets of things.

If x is the name of a thing and A is the name of a property, then $A(x)$ or simply Ax will be used to express the proposition that the property denoted by A is present in the thing denoted by x.

For the quantification of relational propositions (sentences) we shall use the traditional quantifiers ('operators') ( ) and $(E\ )$.

When ( ) or $(E\ )$ is prefixed to a sentence, the blank is filled by a name which occurs in the sentence. It is convenient to say that the sentence is quantified in that name.

The sentence $(x)\ Ax$ is identical with $U\ A$ and the sentence $(Ex)Ax$ is identical with $E\ A$.

Thus the quantifiers ( ) and $(E\ )$ can always replace the quantifiers U and E respectively. The latter quantifiers, however, can replace the former quantifiers, not only before sentences expressing the presence of a property in a thing, but also before sentences expressing the subsistence of a relation between a number of things.

The replacement of the quantifiers ( ) and $(E\ )$ by the quantifiers U and E and the related problems of formal truth, identity, and entailment in sentences involving the symbols ( ) and $(E\ )$ need not be discussed here.

4. On Numbers

The name of a number is a numeral.

Let p and q be names of real numbers.

$-p$ is called the negative numeral of p, and $p+q$ the sum-, p-q the difference-, $p\cdot q$ the product-, and $p:q$ the quotient-numerals of p and q.

A numeral which is neither the negative numeral of another numeral, nor the sum-, difference-, product- or quotient-numeral of two other numerals is called an atomic numeral.

By a molecular complex of n names of real numbers we understand:

i. Any one of the n numerals themselves and their negative numerals.

ii. The sum-, difference-, product- or quotient-numeral of any two of the n numerals.

iii. The negative numeral of any molecular complex of the
n numerals, and the sum-, difference-, product- or
quotient-numeral of any two molecular complexes of
the n numerals.

The n numerals are called the constituents of their molecular
complexes. If the numerals are atomic, they are called atomic
constituents.

Any molecular complex of n names of real numbers denotes
a real number. Which real number it denotes can be calculated
from its atomic constituents.

(The only exception is the quotient-numeral of two numerals,
both of which denote zero.)

If two numerals denote the same number, they are called
identical. It is sometimes convenient to say that identical
numerals denote identical numbers.

If p and q are numerals, the proposition that they are identical
will be expressed $p=q$.

As to the use of brackets we adopt the convention that the
symbol $+$ has a weaker binding force than $-$, $:$, and $\cdot$; the symbol
$-$ than $:$ and $\cdot$; and the symbol $:$ than $\cdot$. Thus, e.g., if p, q, r, s,
and t are numerals, we can instead of $p+(q-(r:(s\cdot t)))$ write
simply $p+q-r:s\cdot t$.

If p and q are numerals, we can for $p\cdot q$ write simply pq.
In virtue of the associative principles we can omit brackets from
sums of sums and products of products. We can henceforth talk
of sums and products of n numbers and of n-termed sum- and
product-numerals respectively.

Let $p_1, \ldots, p_n$ be the names of n real numbers. Then $\sum\limits_{m=1}^{n} p_m$
or simply Σp_m is the name of their sum, and $\prod\limits_{m=1}^{n} p_m$ or simply Πp_m
of their product.

$<$ is the symbol for being smaller than and $>$ the symbol for
being greater than.

Let p and q and δ be names of real numbers. Then the
sentence $p\leqslant q$ is identical with $(p<q)\text{v}(p=q)$ and the sentence
$p\geqslant q$ with $(p>q)\text{v}(p=q)$. Further, the sentence $p=q\pm\delta$ is
identical with $(p\leqslant q+\delta)\&(p\geqslant q-\delta)\&(\delta\geqslant 0)$.

We decide to reserve the Latin letters g, h, i,j, k, l, m, n, u, v
and w and the Greek letters μ and ν for cardinal numbers.

Similarly, we decide to reserve p, q, r, s, t, δ, and ϵ for real numbers (usually in the interval from o and 1 inclusive).

5. On Sequences

Properties may be divided into denumerable and non-denumerable.

Non-denumerable properties are not treated in this inquiry.

That a property is denumerable means that its positive instances can be counted.— That the positive instances can be counted again means that a dyadic relation of a certain kind can be established between the positive instances and cardinal numbers. Such a relation we shall call a *way of counting* (the positive instances of) the property.

A dyadic relation (let us call it R), is a way of counting the positive instances of a property, (let us call it H), if, and only if, the following three conditions are fulfilled:

i. Anything, which is a positive instance of the property (denoted) H, is related by the relation (denoted) R, to a cardinal number.

ii. If a positive instance of H is related by R to a cardinal, then it is not related by R to any other cardinal.

iii. If a positive instance of H is related by R to a cardinal, then no other positive instance of H is related by R to the same cardinal.

In addition to i–iii, let two more conditions be fulfilled:

iv. For the cardinals (denoted) m and n, of which the former is not greater than the latter, and for any cardinal greater than m and smaller than n there is a positive instance of H related to it by R.

v. For no cardinal smaller than m or greater than n above is there a positive instance of H related to it by R.

If there is a relation R and a property H satisfying the conditions i–v, then R is called a dense way of counting H and H is said to be finite. $m+1\text{-}n$ is said to denote the cardinal number of H. In symbols Nc $(H, m+1\text{-}n)$.

If, in iv and v and their symbolic expressions, all reference to the cardinal called n is suppressed, we get two modified conditions iv' and v'. If there is a relation R and a property H satisfying the conditions i–v', then R is a dense way of counting H and H is said to be denumerably infinite.

Two ways, R and R', of counting H are called different, if it is not the case that all positive instances of H are related to the same cardinal number by R and by R'.

If R is a way of counting the property denoted $Hv\overline{H}$, then R is also a way of counting any other property from the same Universe of Properties.

By the n first positive instances of H, when H is counted in the way R, we shall mean the positive instances of H related by R to cardinals not greater than n. It is clear that this definition corresponds to the ordinary use of language only if R is a dense way of counting H, and if the smallest cardinal to which there corresponds a positive instance of H is 1. For technical purposes, however, the more comprehensive use will be convenient.

* * * * *

Any property H and relation R, which satisfy the conditions i–iii above, are said to constitute a sequence. We shall use H,R as a symbol for the sequence.

(This is a particular case of the general concept of a sequence which, however, we do not need here.)

If R is a way of counting H, then R can be used for denumerating, i.e., naming by means of numerals, the positive instances of H. Thus we may introduce the name x_1 for the positive instance of H, which is related by R to the cardinal 1, x_2 for the positive instance related to the cardinal 2, and so on.

If R is a dense way of counting H, beginning from 1, then we shall sometimes, when the names of the property and the way of counting are irrelevant in the context, symbolise the sequence by means of a row of names of numerated things, e.g., $x_1, \ldots, x_n, \ldots$

According to whether the property H is finite or (denumerably) infinite, we shall call the sequence H,R finite or (denumerably) infinite.

As is well known, an (infinite) sequence $p_1, \ldots, p_n, \ldots$ of real numbers is said to converge towards or to approach as its limit the real number p, if for any quantity δ, however small but greater than o, there exists a cardinal m such that for all greater cardinals n the absolute amount of the difference between p and p_n is smaller than δ. In symbols:

(1) $(\delta)\ (Em)\ (n)\ (\delta>0 \& n>m \to p_n = p \pm \delta)$.

It is often convenient to say that the number p_n approaches p as its limit.

The cardinal m is called a point of convergence (towards p, associated with δ).

As an abbreviation for (1) we shall use $lim\ (p_n, p)$.

A sequence $p_1, \ldots, p_n, \ldots$ of real numbers is said to have a partial limit (*limes partialis*) at p, if it is the case that p_n 'again and again' comes 'infinitely near' to p. In symbols:

(2) $(\delta)\ (m)\ (En)\ (\delta>0 \to n>m \& p_n = p \pm \delta)$.

The cardinal n is called a point of condensation.

As an abbreviation for (2) we shall use $plim(p_n, p)$ ·

Of particular interest to us are sequences of real numbers in the interval from o to 1 inclusive.

According to the Bolzano-Weierstrass Principle any sequence of real numbers in the interval from o to 1 (or any other closed interval) *must* have at least one partial limit.

Any sequence in the interval from o to 1 (or any other closed interval) *may* have more than one partial limit.

If a sequence of real numbers $p_1, \ldots, p_n, \ldots$ in the interval from o to 1 has a partial limit at p and no other partial limit, then it approaches p as a limit. This can easily be shown as follows:

Suppose that p were not the limit. This assumption, the denial of (1), means that there would exist an interval round p such that p_n will again and again fall outside this interval (including its limits). In symbols:

(3) $(E\delta)\ (m)\ (En)\ (\delta>0 \& n>m \& \overline{p_n = p \pm \delta})$.

But for numbers between o and 1 inclusive, $p_n = p \pm \delta$ entails $0 \leqslant p_n < p\text{-}\delta \text{ v } p+\delta < p_n \leqslant 1$, meaning that p_n will again and again fall inside either the interval from o to $p\text{-}\delta$ or the interval from $p+\delta$ to 1. But from the Bolzano-Weierstrass Principle it then follows that there is a partial limit within either or both of those intervals. This, however, is contrary to the datum that p is the sole partial limit. It follows by contraposition that p as sole partial limit is also the limit.

That a sequence of real numbers $p_1, \ldots, p_n, \ldots$ in the interval from 0 to 1 inclusive converges towards p, could also be expressed according to the above, by denying the existence of a partial limit at any other value than p. In symbols:

(4) $(q) (E\delta) (Em) (n) (\overline{q=p} \to \delta > 0 \& \overline{(n > m \to p_n = q \pm \delta)})$.

m is here called a point of divergence (from q).

*   *   *   *   *

Let there be a property A and a sequence of properties $A_1, \ldots, A_n, \ldots$.

Let the following three conditions be fulfilled:

   i.   Any property of the sequence is included in its successor property. In symbols:

(5) $(n) (A_n \subset A_{n+1})$.

   ii.   Any positive instance of the property A is also a positive instance of at least one property from the sequence of properties $A_1, \ldots, A_n, \ldots$. In symbols:

(6) $(x) (Ax \to (En) A_n x)$.

   iii.   Anything, which is a positive instance of at least one property from the sequence of properties $A_1, \ldots, A_n, \ldots$, is also a positive instance of the property A. In symbols:

(7) $(x) ((En) A_n x \to Ax)$.

If these three conditions are fulfilled, we say that the sequence of properties $A_1, \ldots, A_n, \ldots$ approaches as its limit the property A.

Let the following three conditions be fulfilled:

   i'.   Any property of the sequence $A_1, \ldots, A_n, \ldots$ (except A_1) is included in its predecessor property. In symbols:

(8) $(n) (A_{n+1} \subset A_n)$.

   ii'.   Any positive instance of the property A is a positive instance of all properties from the sequence $A_1, \ldots, A_n, \ldots$. In symbols:

(9) $(x) (Ax \to (n) A_n x)$.

   iii'.   Anything, which is a positive instance of all properties from the sequence $A_1, \ldots, A_n, \ldots$, is a positive instance of the property A. In symbols:

(10) $(x) ((n) A_n x \to Ax)$.

54

If these three conditions are fulfilled, we also say that the sequence of properties $A_1, \ldots, A_n, \ldots$ approaches as its limit the property A.

That $A_1, \ldots, A_n, \ldots$ approaches A as its limit, will be denoted $lim(A_n, A)$. Thus the sentence $lim(A_n, A)$ is identical with the disjunction-sentence of the conjunction-sentence of (5) and (6) and (7) and the conjunction-sentence of (8) and (9) and (10).

The following illustrations to the two cases, when a sequence of properties is said to approach a certain property as its limit, may be useful to consider:

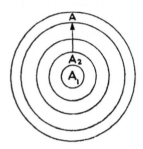 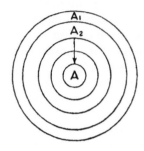

From the left-hand picture we can immediately read off the following important facts: For every n the product $A_n \& A_{n+1}$ is identical with A_n. For every n the product $A_n \& A$ is identical with A_n.

From the right-hand picture again we read off: For every n the product $A_n \& A_{n+1}$ is identical with A_{n+1}. For every n the product $A_n \& A$ is identical with A.

*   *   *   *   *

Let[1] there be a sequence of properties $A_1, \ldots, A_n, \ldots$ (from the same Universe of Properties).

Then there is also a sequence of negation-properties $\overline{A}_1, \ldots, \overline{A}_n, \ldots$.

We shall call the first sequence the positive and the second the negative A-sequence. We may also say that both sequences

[1] The reading of the rest of this paragraph may be postponed until after Chap. VII, §12.

are made up of the same properties, taken positively and negatively respectively. (This mode of speech is, of course, logically objectionable, but sometimes convenient to use.)

A conjunction (disjunction) of properties from the positive A-sequence we call, accordingly, a positive conjunction (disjunction), and a conjunction (disjunction) from the negative A-sequence we call a negative conjunction (disjunction). A conjunction (disjunction) of properties, some of which belong to one and some to the other of the two sequences, we call a mixed conjunction (disjunction).

It will be convenient to regard the members of the two A-sequences themselves as positive and negative conjunctions (disjunctions) of one single constituent only.

A mixed conjunction is called consistent or not-contradictory, if none of its constituents is the negation of another of its constituents. A positive and a negative conjunction are said to contradict one another, if at least one of the constituents of the first conjunction is the negation of a constituent of the second conjunction, or *vice versa*.

By the A-constituents of a positive or negative conjunction (disjunction) we understand the *positive A-properties* of which (or of the negation-properties of which) it is a conjunction (disjunction).

Consider the arrangement of the members of the positive A-sequence into groups, according to the following scheme, the constructive principle of which should be obvious to the reader:

Scheme I

A_1,
A_2, A_1A_2,
A_3, A_1A_3, A_2A_3, $A_1A_2A_3$,
A_4, A_1A_4, A_2A_4, A_3A_4, $A_1A_2A_4$, $A_1A_3A_4$, $A_2A_3A_4$, $A_1A_2A_3A_4$,

---

By inserting conjunctions we generate from this scheme a sequence of properties A_1, A_2, $A_1 \& A_2$, A_3, $A_1 \& A_3$, $A_2 \& A_3$, ... This sequence contains only the positive conjunctions. For these conjunctions we introduce in order the new names $K_1, \ldots, K_n, \ldots$. Thus (the name) K_1 is identical with (denotes the same property as the name) A_1, K_5 is identical with $A_1 \& A_3$, *etc.* We refer to this sequence as the (one-dimensional) K-sequence.

By inserting disjunctions we generate from this scheme a sequence of properties A_1, A_2, $A_1 \vee A_2$, A_3, $A_1 \vee A_3$, $A_2 \vee A_3$, ... This sequence contains only the positive disjunctions. For these disjunctions we introduce in order the new names M_1, ..., M_n, ... and refer to the sequence as the (one-dimensional) M-sequence.

If we take the negation-properties of the members of each group and, provided the group has more than one member, join the negation-properties by conjunction, we get a sequence of properties $\overline{A_1}$, $\overline{A_2}$, $\overline{A_1} \& \overline{A_2}$, $\overline{A_3}$, $\overline{A_1} \& \overline{A_3}$, $\overline{A_2} \& \overline{A_3}$, ... This sequence contains only negative conjunctions. For these conjunctions we introduce in order the new names L_1, ...,L_n, ... and refer to the sequence as the (one-dimensional) L-sequence.

(The sequence of all negative disjunctions will not be needed.)

It would be possible to give a rule for the calculation, for any given value of n, of the number of A-constituents of the conjunctions K_n, L_n and the disjunction M_n respectively. It may, however, be regarded as inconvenient that the number of such constituents cannot be directly read off from the indices of the names of the respective members of the K-, L-, and M-sequences.

In order to do away with this inconvenience, we rearrange the groups of *Scheme I* so that we get:

Scheme II

A_1, A_2, A_3, A_4, ...

$A_1 A_2$, $A_1 A_3$, $A_2 A_3$, $A_1 A_4$, $A_2 A_4$, $A_3 A_4$, ...

$A_1 A_2 A_3$, $A_1 A_2 A_4$, $A_1 A_3 A_4$, $A_2 A_3 A_4$, ...

$A_1 A_2 A_3 A_4$, ...

(The principle of rearrangement should be obvious to the reader.)

We can now split up the one-dimensional K-, L-, and M-sequences into two-dimensional K-, L-, and M-sequences:

$$^1K_1, \ldots, {}^nK_1, \ldots \qquad {}^1L_1, \ldots, {}^nL_1, \ldots \qquad {}^1M_1, \ldots, {}^nM_1, \ldots$$

$$^1K_m, \ldots, {}^nK_m, \ldots \quad {}^1L_m, \ldots, {}^nL_m, \ldots \quad {}^1M_m, \ldots, {}^nM_m, \ldots$$

Here (the names) $^1K_1$, $^2K_1$, *etc.* are identical with (the names) A_1, A_2, *etc.*; (the names) $^1K_2$, $^2K_2$, *etc.* with (the names) $A_1 \& A_2$, $A_1 \& A_3$, *etc.*; (the names) $^1K_3$, $^2K_3$, *etc.* with (the names) $A_1 \& A_2 \& A_3$, $A_1 \& A_2 \& A_4$, *etc.*; and so on.

Further $^1L_1$, $^2L_1$, *etc.* are identical with $\bar{A}_1$, $\bar{A}_2$, *etc.*; $^1L_2$, $^2L_2$, *etc.* with $\bar{A}_1 \& \bar{A}_2$, $\bar{A}_1 \& \bar{A}_3$, *etc.*; and so on.

Finally $^1M_1$, $^2M_1$, *etc.* are identical with A_1, A_2, *etc.*; $^1M_2$, $^2M_2$, *etc.* with $A_1 \vee A_2$, $A_1 \vee A_3$, *etc.*; and so on.

Thus ^nK_m is the conjunction, and ^nM_m the disjunction of the same m properties from the positive A-sequence. ^nL_m again is the conjunction of the corresponding properties from the negative A-sequence. $^nK_1$ and $^nM_1$ are both identical with A_n, and $^nL_1$ is identical with $\bar{A}_n$.

In the two-dimensional K-, L-, and M- sequences the number of A-constituents of the conjunctions and disjunctions respectively can be directly read off from the lower indices of their names.

Given i and n, the A-constituents of the conjunction ^iK_n are uniquely determined. m of those n constituents are selected. The selection can take place in $\binom{n}{m}$ different ways. Consider the conjunctions of the properties thus selected. Consider then the disjunction of all these $\binom{n}{m}$ conjunctions. Its A-constituents are uniquely determined, given i, n, and m. We call the disjunction of these conjunctions $_m^iQ^n$.

Similarly, given i and n, the A-constituents of the conjunction ^iL_n are uniquely determined. m of those n constituents are selected. The selection can again take place in $\binom{n}{m}$ different ways. Consider the conjunctions of the negations of the properties thus selected. Consider then the disjunction of all these $\binom{n}{m}$ conjunctions. Its A-constituents are uniquely determined, given i, n, and m. We call the disjunction of these conjunctions $_m^iR^n$,

(It is easy to see, though of no particular importance, that ${}_{1}^{i}Q^{n}$ is the same as ${}^{i}M_{n}$. For consider what presence-functions of A-constituents they represent; ${}^{i}M_{n}$ is a disjunction of n members from the positive A-sequence. ${}_{1}^{i}Q^{n}$ again is the disjunction of those n members from the positive A-sequence, the conjunction of which is ${}^{i}K_{n}$. But, as was observed above, the A-constituents of ${}^{i}K_{n}$ are the same as those of ${}^{i}M_{n}$. Therefore ${}_{1}^{i}Q^{n}$ and ${}^{i}M_{n}$ are disjunctions of the same A-constituents.)

Consider the property ${}_{m}^{i}Q^{n}\&_{n-m}{}^{i}R^{n}$. We leave it as an exercise to the reader to satisfy himself that this property is the disjunction of all the $\binom{n}{m}$ mixed conjunctions which can be formed of n given A-constituents by taking m of them positively and the remaining n-m ones negatively.

For the disjunction, of all the $\binom{n}{m_1}$ mixed conjunctions which can be formed of n given A-constituents by taking m_1 of them positively and the remaining n-m_1 ones negatively, and of the $\binom{n}{m_2}$ mixed conjunctions which can be formed of n given A-constituents by taking m_2 of them positively and the remaining n-m_2 ones negatively, and . . ., and of the $\binom{n}{m_k}$ mixed conjunctions which can be formed of n given A-constituents by taking m_k of them positively and the remaining n-m_k ones negatively, —we introduce the name ${}_{m_1,\overset{i}{\ldots},m_k}\triangle^{n}$.

Let us next consider two groups of properties when grouped according to *Scheme II* above; say, the (group of properties named by the names in the) k:th group of the g:th horizontal row and the n:th group of the m:th horizontal row. Let us assume that $g \leqslant m$. It is now possible to express, in terms of arithmetical conditions for indices, what it means to say that all properties in the first group also occur in the second group. Any one of the g indices in the first group may be identical with some one of the m indices in the second group. Thus there are in all gm equations or identity-sentences, expressing identity of indices. That every index in the first group is identical with some index in the second group means that of those gm equations g are simultaneously valid. This again is possible in $m!{:}g!$ different ways, since all indices in the same group are different and consequently can be identical with one index only in the other group. The proposition that all properties in the first group also occur in the second group is thus expressed by an $m!{:}g!$-termed disjunction-sentence of g-termed conjunction-sentences of identity-sentences between indices. The indices are uniquely determined, given g, k, m, and n. For the proposition in question we shall introduce the sentence $Inc\,(n, m, k, g)$.

Considering the meaning of the expressions, it is easy to see that $Inc\,(n, m, k, g)$ entails $^kM_g \subset {}^nM_m$.

Similarly, it is easy to see that $Inc\,(n, m, k, g)$ entails $^nK_m \subset {}^kK_g$ and also $^nL_m \subset {}^kL_g$.

The conjunction-sentence of the negation-sentences of the gm identity-sentences above, expresses the proposition that no property (named) in the k:th group of the g:th horizontal row of *Scheme II* also occurs in the n:th group of the m:th horizontal row. For this proposition we introduce the sentence $Exc\,(n, m, k, g)$.

It follows that $\overline{Exc}\,(n, m, k, g)$ means that at least one property (named) in the k:th group of the g:th horizontal row of *Scheme II* also occurs in the n:th group of the m:th horizontal row.

Considering the meaning of the expressions, it is easy to see that $\overline{Exc}\,(n, m, k, g)$ means that the mixed conjunctions ^kK_g & ^nL_m and ^nK_m & ^kL_g are contradictory. $Exc\,(n, m, k, g)$ consequently means that the mixed conjunctions in question are not-contradictory or consistent.

Similarly, $\overline{Exc}\,(n, m, k, g)$ entails that the positive conjunctions ^kK_g and ^nK_m respectively contradict the negative conjunctions ^nL_m and ^kL_g respectively. $Exc\,(n, m, k, g)$ entails that the positive and negative conjunctions in question do not contradict each other.

Let there be a sequence $p_1, \ldots, p_n, \ldots$ of real numbers between o and 1 inclusive.

From this sequence we can derive another sequence $1-p_1, \ldots, 1-p_n, \ldots$ of real numbers between o and 1 inclusive.

We shall call the sequences the p- and the $1-p$-sequences respectively.

The products and sums of members of the p- and $1-p$-sequences respectively can be arranged into sequences in accordance with schemes, which correspond to *Scheme I* and *Scheme II* above for the arrangement of products and sums of members of the positive and negative A-sequences respectively.

Thus we get a sequence of products $\pi_1, \ldots, \pi_n, \ldots$ of

p- numbers corresponding to the one-dimensional K-sequence, a sequence of products $\rho_1, \ldots, \rho_n, \ldots$ of 1-p-numbers corresponding to the one-dimensional L-sequence, and a sequence of sums $\sigma_1, \ldots, \sigma_n, \ldots$ of p-numbers corresponding to the one-dimensional M-sequence.

These sequences may in their turn be split up into two-dimensional sequences:

$^1\pi_1, \ldots, \,^n\pi_1, \ldots$	$^1\rho_1, \ldots, \,^n\rho_1, \ldots$	$^1\sigma_1, \ldots, \,^n\sigma_1, \ldots$
$^1\pi_m, \ldots, \,^n\pi_m, \ldots$	$^1\rho_m, \ldots, \,^n\rho_m, \ldots$	$^1\sigma_m, \ldots, \,^n\sigma_m, \ldots$

corresponding to the two-dimensional K-, L-, and M-sequences.

It should be noted that $^n\pi_1$ and $^n\sigma_1$ are both identical with p_n, and that $^n\rho_1$ is identical with 1-p_n.

From $^n\pi_m$ we derive $^n_i\Pi^m$ in the same way that we derived $^n_i Q^m$ from $^n K_m$.

The Π-numbers are sums of products of p-numbers. (Sums of products of 1-p-numbers are not needed.)

It should be observed that, just as $^n_i Q^m$ is identical with $^n M_m$, so $^n_i\Pi^m$ is identical with $^n\sigma_m$. (Cf. above p. 59.)

Let there be a two-dimensional sequence of real numbers between o and 1 inclusive:

$$^1p_1, \ldots, \,^n p_1, \ldots$$

$$^1p_m, \ldots, \,^n p_m, \ldots$$

Consider the one-one-correspondence between this sequence and the two-dimensional K-sequence, which makes $^1 K_1$ answer to $^1 p_1$, etc. Then to $^n_i Q^m$ there will answer a sum $^n_i\theta^m$ of members of this two-dimensional p-sequence.

If the respective members of the sequence $p_1, \ldots, p_n, \ldots$ happened to be identical in order with the respective members of the sequence $^1p_1, \ldots, \,^n p_1, \ldots$, then $^n_1\Pi^m$ would be identical with $^n_1\theta^m$.

Let there, finally, be two one-dimensional sequences of real numbers between 0 and 1 inclusive: $p_1, \ldots, p_n, \ldots$ and $q_1, \ldots, q_n, \ldots$.

For $p_1 q_1$ we introduce the new name w_1, and $\ldots$, and for $p_n q_n$ the new name w_n.

For the sum of all w-numbers with the same indices as the A-constitutents of ^nM_n we introduce the name $_n\Omega_m$.

Chapter Three

THE FORM OF INDUCTIVE ARGUMENTS
LAWS OF NATURE

1. Induction as Inference

WE are now in a position which enables us to describe our object of study, loosely outlined in Chap. I, in the more precise terms of logic which were introduced in Chap. II.

As already observed (p. 20), one of the essential characteristics of inductive inference is its inconclusive or non-demonstrative nature. This means that the conclusion of the argument never follows from the (conjunction of the) premisses.

There is a type of conclusive or demonstrative argument which bears a superficial resemblance to inductive inference. It is sometimes called Complete, Perfect, or Non-Problematic Induction and contrasted with the inconclusive or non-demonstrative type of argument, which is then called Incomplete, Imperfect, or Problematic Induction.

Inductive inference in the genuine sense should, of course, be distinguished also from certain types of demonstrative argument in logic and mathematics, which are called 'inductive'; such as, *e.g.*, so-called Bernoullian Induction, or reasoning from n to $n+1$.

In replacing our previous rough description of inductive inference by a more exact one, it is convenient to start from induction of the second order, or induction leading to theories (as opposed to predictions).

The conclusion of an inductive inference of the second order is a Universal Implication or Equivalence.

A theory is thus a Universal Implication or Equivalence.

63

As we have seen (p. 43 f.), a Universal Implication or Equivalence is a proposition to the effect that an implication-property or equivalence-property is universal. If the property in question is (denumerably) infinite, the theory is said to have the numerically unrestricted range of application (p.15) which entitles it to the name of a law (Law of Nature).

The premisses of an inductive inference of the second order are confirming instances of the conclusion, *i.e.* the Universal Implication or Equivalence. This definition of the premisses is considerably more comprehensive than the one given (p. 15) for the purpose of a first approximation, as will be seen from the following considerations.

A confirming instance of a Universal Implication is a true proposition to the effect that a thing is a positive instance of the implication-property under consideration.

It should be observed that one and the same property can be the implication-property of different antecedents and consequents. *E.g.*, the implication-property of the property (called) *A* as antecedent and the property (called) *B* as consequent is the same as the implication-property of $\overline{B}$ as antecedent and $\overline{A}$ as consequent.

Given an antecedent and a consequent and a Universal Implication it follows from the structure of the implication-property that there are three types of confirming instances of the Universal Implication:

i. Confirmations afforded (p. 44) by things, in which both the antecedent and the consequent of the implication-property are present.

ii. Confirmations afforded by things, in which the antecedent is absent and the consequent present.

iii. Confirmations afforded by things, in which both the antecedent and the consequent are absent.

It can be seen that any negative instance of the antecedent is a positive instance of the implication-property, irrespective of whether it is a positive or a negative instance of the consequent. The same is true, *mutatis mutandis*, of any positive instance of the consequent. Thus anything, which is known to be a negative instance of the antecedent or a positive instance of the consequent, is known as such to afford a confirming instance of the Universal Implication. This fact, which has startled some philosophers, we shall refer to as the Paradox of Confirmation.

A confirming instance of a Universal Equivalence is a true proposition to the effect that a thing is a positive instance of the equivalence-property under consideration. From the structure of the equivalence-property it follows that there are two types of confirming instances of Universal Equivalences, corresponding *mutatis mutandis* to i and iii above. It further follows that the Paradox of Confirmation does not arise in connexion with Universal Equivalences.

We now turn to induction of the first order.

By the test-conditions of a Universal Implication, given an antecedent and a consequent, we shall understand propositions to the effect that the antecedent is present (first test-condition), or absent (second test-condition); or that the consequent is present (third test-condition), or absent (fourth test-condition), in a thing. Similarly, we define *mutatis mutandis* the four test-conditions of a Universal Equivalence.

From the Universal Implication and its first test-condition we can deduce its third test-condition, *i.e.*, from the conjunction of the Universal Implication and the first test-condition the third test-condition follows. From the Universal Implication and its fourth test-condition we can again deduce its second test-condition. But from the Universal Implication in conjunction with its second or third test-condition none of its other test-conditions can be deduced. We, therefore, decide to call the first and the fourth test-condition of a Universal Implication real, and the second and third apparent.

From the Universal Equivalence and its first test-condition we can deduce its third test-condition, and *vice versa*. From the Universal Equivalence and its second test-condition we can deduce its fourth test-condition, and *vice versa*. Thus all the test-conditions of a Universal Equivalence are real.

The conclusion of an inductive inference of the first order is a test-condition of a certain Universal Implication or Equivalence.

The premisses of an inductive inference of the first order are:

i. Confirming instances of this same Universal Implication or Equivalence, and :

ii. A test-condition of it, from which, in conjunction with the Universal Implication or Equivalence, that test-condition which is the conclusion can be deduced. (Cf. above p. 15).

It should be observed that the Universal Implication or Equivalence is only 'imaginary': it does not enter into the argument.

It should be clear from the above considerations, in what sense induction of the second order can be regarded as logically more 'basic' than induction of the first order.

Throughout this book it will be assumed that the conclusions of inductive arguments of the second order are theories in the stronger sense of laws. This limitation is here, on the whole, of technical importance only.

2. The Logic of Conditions

Having briefly examined the form of inductive inferences as wholes, we shall now turn to a closer examination of the form of inductive conclusions. We shall here be concerned only with conclusions of inductive arguments of the second order, *i.e.*, with Universal Implications and Equivalences.

D1. That (the property denoted by) A is a Sufficient Condition of (the property denoted by) B means that whenever A is present, then B is also present, or that $A \subset B$.

D2. That (the property denoted by) A is a Necessary Condition of (the property denoted by) B means that whenever B is present, then A is also present, or that $B \subset A$.

D3. That (the property denoted by) A is a Necessary-and-Sufficient Condition of (the property denoted by) B means that whenever, and only whenever, A is present, then B is also present, or that $A \equiv B$.

A Universal Implication or Equivalence thus establishes a connexion of condition between properties ('terms,' 'factors').

A connexion of condition between properties may also be called a connexion of law or nomic connexion.

On the basis of D_1-D_3 and some additional definitions to be given later, it is possible to prove a number of theorems concerning Sufficient, Necessary, and Necessary-and-Sufficient Conditions. The system constituted by the definitions and theorems, may be called the Logic of Conditions.

(The Logic of Conditions is thus the logic of nomic connexion.)

T_1. If A is a Sufficient Condition of B, then B is a Necessary Condition of A. The converse of this is also true.

The theorem follows immediately from D_1 and D_2.

T_2. The following four propositions are identical, *viz.*

i. that the presence of A is sufficient for the presence of B,
ii. that the absence of A is necessary for the absence of B,
iii. that the absence of B is sufficient for the absence of A, and
iv. that the presence of B is necessary for the presence of A.

Proof: $A \subset B$ is identical with $\bar{B} \subset \bar{A}$. It follows from T_1 that each one of the identical sentences can be read in two different ways as expressing a relation of condition. Thus there are four different ways of reading in all. These are the ways mentioned in the theorem.

T_3. That A is a Necessary-and-Sufficient Condition of B can be expressed in sixteen different ways in terms of Sufficient and Necessary Conditions. Of those sixteen ways we mention the following four, *viz.*,

i. that the presence of A is sufficient for the presence of B and the absence of A for the absence of B,
ii. that the presence of A is sufficient as well as necessary for the presence of B,
iii. that the absence of A is necessary for the absence of B and the presence of A for the presence of B, and
iv. that the absence of A is necessary as well as sufficient for the absence of B.

Proof: $A \equiv B$ is identical with i. $(A \subset B) \& (\bar{A} \subset \bar{B})$, ii. $(A \subset B) \& (B \subset A)$, iii. $(\bar{B} \subset \bar{A}) \& (B \subset A)$, and iv. $(\bar{B} \subset \bar{A}) \& (\bar{A} \subset \bar{B})$.

It follows from $T1$ that each one of the constituents of the conjunction-sentences can be read in two different ways. Thus each one of the conjunction-sentences themselves can be read in four different ways, and there are in all four times four, or sixteen, different ways of reading. Among them will be found the four ways mentioned in the theorem.

$T4$. The relation of being a Necessary-and-Sufficient Condition is reflexive, symmetrical, and transitive.

Proof: $A \equiv A$ is tautologous. $A \equiv B$ is identical with $B \equiv A$. $(A \equiv B)\&(B \equiv C)$ entails $A \equiv C$.

$T5$. The relation of being a Sufficient Condition is reflexive and transitive, but not symmetrical.

Proof: $A \subset A$ is tautologous. $(A \subset B)\&(B \subset C)$ entails $A \subset C$. $A \subset B$ is not identical with $B \subset A$.

$T6$. The relation of being a Necessary Condition is also reflexive and transitive, but not symmetrical.

Proof: $A \subset A$ is tautologous. $(B \subset A)\&(C \subset B)$ entails $C \subset A$. $A \subset B$ is not identical with $B \subset A$.

A property may have more than one Sufficient or Necessary Condition. If this is the case, we speak of Plurality of Conditions.

$T7$. Let a property be the sum of n properties. If this property is a Sufficient Condition of a given property, then every one of the n properties is also a Sufficient Condition of the given property. Conversely, if every one of the n properties is a Sufficient Condition of a given property, then their sum is also a Sufficient Condition of the given property.

Proof: $AvB \subset C$ is identical with $(A \subset C)\&(B \subset C)$.

$T8$. Let a property be the product of n properties. If this property is a Necessary Condition of a given property, then every one of the n properties is also a Necessary Condition of the given property. Conversely, if every one of the n properties is a Necessary Condition of a given property, then their product is also a Necessary Condition of the given property.

Proof: $A \subset B\&C$ is identical with $(A \subset B)\&(A \subset C)$.

$T9$. Let a Sufficient (Necessary) Condition of a given property be a presence-function of n properties. Such a Sufficient (Necessary) Condition is, in general, the sum (product) of a plurality of Sufficient (Necessary) Conditions of the given property, every one of which is the product (sum) of m of the n

properties and of the negation-properties of the remaining n-m properties.

Proof: The theorem follows immediately from $T7$ ($T8$) and from the fact that the name of the Sufficient (Necessary) Condition of the given property has a perfect disjunctive (conjunctive) normal form in terms of the names of the n properties.

The qualification 'in general' means 'unless the Sufficient (Necessary) Condition happen to be the contradiction (tautology) or product (sum) of the n properties.'

The status of Necessary-and-Sufficient Conditions with regard to plurality is somewhat different from that of Sufficient and Necessary Conditions.

$T10$. If a property has several Necessary-and-Sufficient Conditions, they are co-extensive, and thus Necessary-and-Sufficient Conditions of one another.

Proof: $(A \equiv B) \& (A \equiv C)$ entails $B \equiv C$.

$T11$. Let a Necessary-and-Sufficient Condition of a given property be the sum (product) of n properties. Then, in general, no one of the n properties is also a Necessary-and-Sufficient Condition of the given property.

Proof: Neither $A \equiv BvC$ nor $A \equiv B \& C$ entails $(A \equiv B)v(A \equiv C)$.

The qualification 'in general' means 'unless all the other n-1 properties happen to be empty (universal).'

$T12$. Let a Necessary-and-Sufficient Condition of a given property be a presence-function of n properties. Such a Necessary-and-Sufficient Condition is not, in general, the sum or product of a plurality of Necessary-and-Sufficient Conditions of the given property, every one of which is the product or sum of m of the n properties and the negation-properties of the remaining n-m properties.

Proof: The theorem immediately follows from $T11$ and considerations as regards the perfect normal forms of the name of the Necessary-and-Sufficient Condition.

The qualification 'in general' means 'unless the given property happens to be empty or universal.'

$T13$. If A is a Necessary-and-Sufficient Condition of B, then the sum of A and any of A's Sufficient Conditions is a Necessary-and-Sufficient Condition of B. So also is the product of A and any of A's Necessary Conditions.

Proof: $(A \equiv B)\&(C \subset A)$ entails $AvC \equiv B$, and $(A \equiv B)\&$ $(A \subset C)$ entails $A\&C \equiv B$.

T14. If A is a Sufficient Condition of B, then the product of A and any property is also a Sufficient Condition of B.

Proof: $A \subset B$ entails $A\&C \subset B$.

T15. If A is a Necessary Condition of B, then the sum of A and any property is also a Necessary Condition of B.

Proof: $B \subset A$ entails $B \subset AvC$.

T16. Any empty property is a Sufficient Condition of any property.

Proof: $\bar{E} A$ entails $A \subset B$.

T17. Any universal property is a Necessary Condition of any property.

Proof: $U A$ entails $B \subset A$.

The following theorems concerning the interrelatedness of Sufficient and Necessary Conditions should be mentioned:

T18. If the absence of at least one of a number of properties is sufficient for the absence of a certain property A, then the presence of all those properties is necessary for the presence of A. And the converse of this holds also.

Proof: $\bar{B}v\bar{C} \subset \bar{A}$ is identical with $A \subset B\&C$.

T19. If the absence of every one of a number of properties is necessary for the absence of a certain property A, then the presence of at least one of those properties is sufficient for the presence of A. And the converse of this holds also.

Proof: $\bar{A} \subset \bar{B}\&\bar{C}$ is identical with $BvC \subset A$.

T20. If the absence of every one of a number of properties is sufficient for the absence of a certain property A, then the presence of at least one of those properties is necessary for the presence of A. And the converse of this is also true.

Proof: $\bar{B}\&\bar{C} \subset \bar{A}$ is identical with $A \subset BvC$.

T21. If the absence of at least one of a number of properties is necessary for the absence of a certain property A, then the presence of all those properties is sufficient for the presence of A. And the converse of this is also true.

Proof: $\bar{A} \subset \bar{B}v\bar{C}$ is identical with $B\&C \subset A$.

The last four theorems serve to illustrate the fact that the negation of a property is just as much a 'property' as the property itself. This is clear from the point of view of logic. It is usually, but not always, evident from the point of view of ordinary language also. Frequently language uses a 'positive' expression for the absence as well as for the presence of a property. The negation, *e.g.*, of the property of having a constant temperature is the property of having a variable temperature.

Again, the last two theorems should make it quite clear, that if conjunction-properties be accepted as Sufficient Conditions (which they obviously must), then disjunction-properties must be accepted as Necessary Conditions (which *prima facie* is perhaps less evident and sometimes even disputed). To give an example: If it be sufficient in keeping the volume of a gas constant to keep the pressure *and* the temperature constant, then it will be necessary in effecting a variation in the volume to let the pressure *or* the temperature be varied.

Let ϕ_0 denote a set of logically totally independent properties.

D4. That A is a Sufficient (Necessary, Necessary-and-Sufficient) Condition of B in ϕ_0 means that A is a Sufficient (Necessary, Necessary-and-Sufficient) Condition of B *and* a presence-function of some n properties in ϕ_0.

D5. That A is a Greatest Sufficient Condition of B in ϕ_0 means that A is a Sufficient Condition of B in ϕ_0 and that no member or conjunction of members of ϕ_0 which includes (but is not included in) A is a Sufficient Condition of B.

D6. That A is a Smallest Necessary Condition of B in ϕ_0 means that A is a Necessary Condition of B in ϕ_0 and that no member or disjunction of members of ϕ_0 which is included in (but does not include) A is a Necessary Condition of B.

D7. That A is the Total Sufficient Condition of B in ϕ_0 means that A is the sum of all Greatest Sufficient Conditions of B in ϕ_0.

D8. That A is the Total Necessary Condition of B in ϕ_0 means that A is the product of all Smallest Necessary Conditions of B in ϕ_0.

It is of no interest to introduce the notions of Greatest, Smallest, and Total Necessary-and-Sufficient Conditions, since the Necessary-and-Sufficient Conditions of a property are all co-extensive ($T10$).

T22. A property's Total Sufficient Condition in ϕ_0 is co-extensive with the sum of all its Sufficient Conditions in ϕ_0.

Proof: Let A be the Total Sufficient Condition of B in ϕ_0. Let C be any Sufficient Condition of B in ϕ_0. C is either a Greatest Sufficient Condition of B in ϕ_0, or not. In the first case it is as such included in A. In the second case it is *per definitionem* included in some Greatest Sufficient Condition of B in ϕ_0 and, since inclusion is transitive, in A. But $C \subset A$ entails $A v C \equiv A$.

T23. A property's Total Necessary Condition in ϕ_0 is co-extensive with the product of all its Necessary Conditions in ϕ_0.

Proof: Let A be the Total Necessary Condition of B in ϕ_0. Let C be any Necessary Condition of B in ϕ_0. C is either a Smallest Necessary Condition of B in ϕ_0, or not. In the first case it as such includes A. In the second case it *per definitionem* includes some Smallest Necessary Condition of B in ϕ_0 and, since inclusion is transitive, also A. But $A \subset C$ entails $A \& C \equiv A$.

T24. Given a property A and a set ϕ_0, in every positive instance of A the Total Necessary Condition of A in ϕ_0 must be present.

This immediately follows from *D2, D4, D6, D8,* and *T8.*

T25. Given a property A and a set ϕ_0, in some positive instance of A the Total Sufficient Condition of A in ϕ_0 may be absent.

This follows from *D1, D4, D5, D7,* and *T5* and *T7.*

The above difference between Sufficient and Necessary Conditions is of importance for the study of inductive inference.

D9. That A is a Determined Property in ϕ_0 means that in every positive instance of A the Total Sufficient Condition of A in ϕ_0 is present.

T26. If A is a Determined Property in ϕ_0, then its Total Sufficient Condition in ϕ_0 is its Necessary-and-Sufficient Condition.

This immediately follows from *D3* and *D9.*

T27. If A is a Determined Property in ϕ_0, then its Total Necessary Condition in ϕ_0 is also its Necessary-and-Sufficient Condition.

Proof: Let B be the Total Sufficient and C the Total Necessary Condition of A in ϕ_0. According to $T26$ we have $A \equiv B$. This entails $A \subset B$. Thus B is a Necessary Condition of A. According to $T23$ we have $C \equiv C\&B$. $(B \subset A)\&(A \subset C)$ entails $B \subset C$. $B \subset C$ entails $B \equiv C\&B$. $(C \equiv C\&B)\&(B \equiv C\&B)$ entails $B \equiv C$. $(A \equiv B)\&(B \equiv C)$ entails $A \equiv C$. Thus C is the Necessary-and-Sufficient Condition of A.

$T28$. If A is a Determined Property in ϕ_0, then its Total Sufficient Condition in ϕ_0 and its Total Necessary Condition in ϕ_0 are co-extensive.

This immediately follows from $T26$ and $T27$.

$D10$. That A is a Contributory Condition of B in ϕ_0 means that A is a Necessary Condition in ϕ_0 of at least one Sufficient Condition of B in ϕ_0.

$D11$. That A is an Indispensable Contributory Condition of B in ϕ_0 means that A is a Necessary Condition in ϕ_0 of all Sufficient Conditions of B in ϕ_0.

$D12$. That A is a Substitutable Requirement of B in ϕ_0 means that A is a Sufficient Condition in ϕ_0 of at least one Necessary Condition of B in ϕ_0.

$D13$. That A is a Counteracting Condition of B in ϕ_0 means that $\bar{A}$ is a Contributory Condition of B in ϕ_0.

$T29$. If A is a Counteracting Condition of B in ϕ_0, then A is a Substitutable Requirement of $\bar{B}$ in ϕ_0.

Proof: $\bar{A}\&C \subset B$ is identical with $\bar{B} \subset A v \bar{C}$.

In order to cut out the trivialities which may arise in virtue of $T16$ and $T17$ we shall henceforth adopt the following convention:

$C1$. By a Sufficient Condition we shall never understand a property known to be empty and by a Necessary Condition we shall never understand a property known to be universal.

In order to cut out the trivialities which may arise in virtue of $T14$ and $T15$ we shall henceforth adopt the following convention:

$C2$. By a Sufficient Condition (in a set ϕ_0) we shall always understand a Greatest Sufficient Condition (in ϕ_0) and by a Necessary Condition (in a set ϕ_0) we shall always understand a Smallest Necessary Condition (in ϕ_0).

In view of $T9$ we shall henceforth adopt the following convention:

C3. By a Sufficient Condition (in a set ϕ_0) we shall always understand a product of m ($0 \leqslant m \leqslant n$) of some n ($1 \leqslant n$) properties (in ϕ_0), and the negation-properties of the remaining n-m properties; and by a Necessary Condition (in a set ϕ_0) we shall always understand a sum of m ($0 \leqslant m \leqslant n$) of some n ($1 \leqslant n$) properties (in ϕ_0), and the negation-properties of the remaining n-m properties.

It should be noted that the introduction of *C3* does not mean any limitation in the scope of our treatment of conditions. It does not follow that Sufficient Conditions which are other presence-functions than products, or that Necessary Conditions which are other presence-functions than sums, of m of some n properties and the negation-properties of the remaining n-m properties, are henceforth omitted from treatment, but simply that their treatment is reducible to cases of Plurality of Conditions in virtue of *T9.*

It should be noted that, in view of *T12*, it is not possible to extend *C3* to Necessary-and-Sufficient Conditions without effecting a real limitation of the scope of inquiry.

A Greatest Sufficient Condition (Smallest Necessary Condition) of A in ϕ_0 which is a one-termed product (sum) we shall call Simple (in ϕ_0). A Greatest Sufficient Condition (Smallest Necessary Condition) of A in ϕ_0 which is a more-than-one-termed product (sum) we shall call Complex (in ϕ_0). It is convenient to let the number of terms of the products (sums) measure the degree of complexity of the condition. 1-complex conditions are simple.

A Necessary-and-Sufficient Condition of A in ϕ_0 which is a presence-function of a single property in ϕ_0 will be called Simple (in ϕ_0). A Necessary-and-Sufficient Condition of A in ϕ_0 which is not Simple (in ϕ_0) will be called Complex (in ϕ_0). It is convenient to let the minimum number of properties in ϕ_0 of which a Necessary-and-Sufficient Condition of A in ϕ_0 is a presence-function, measure its degree of complexity.

It is to be observed that the conditioning relation, as defined here, has nothing to do with relatedness in time. If, for instance, it is admitted that rain is a Sufficient Condition of the ground becoming wet, it follows that the ground becoming wet is a Necessary Condition of rain. This, however, may appear awkward from the point of view of ordinary language. The difficulty raised by ordinary language indicates that our popular notions of conditioning are influenced by our popular notions of causality.

What is the connexion between the conditioning relation and causality? It is not unplausible to assume that the notion of condition could be used for what might be called a partial analysis of the notions of cause and effect. By this we mean that causal relationship would imply a conditional relationship, *i.e.*, a nomic connexion, though not conversely.

If the converse were also the case, the analysis might be called total. It is not unplausible to think that what is required in order to make the suggested partial analysis of causal relationship total, is at least the introduction of certain qualifications as regards relatedness in time. Everyone seems to think that the effect cannot come into existence before the cause, and most people are further inclined to hold that the cause must come into existence before the effect.

If a partial analysis of causality in terms of conditions is accepted, one might, in conformity with our definitions above, distinguish between sufficient cause, necessary cause, necessary-and-sufficient cause, contributory cause, indispensable contributory cause, substitutable cause, and counteracting cause. To what extent similar distinctions apply to the notion of effect will not be considered here.

Further, if a partial analysis of causality in terms of condition is accepted, it is plausible to assume that the notion of a Determined Property will throw some light upon what philosophers, scientists, and ordinary people mainly have in mind when talking vaguely of phenomena as being causally determined.

It may be suggested that the principle 'nothing occurs without a cause' should be understood as implying that all properties are Determined Properties. We shall refer to this principle, without pretending to historical adequacy, as the Principle of Determinism.

It should, however, be observed that the notion of a Determined Property, as defined by us, is a relative notion. It is relative to the selection of a set of properties ϕ_0. There is, moreover, no set embracing 'all' properties, since the selection of the set is always confined to a certain Universe of Properties. (Cf. Chap. II, §2.)

It is fairly obvious that the Principle of Determinism cannot be claimed to apply to all Universes of Properties. It hardly applies to the various universes of what may be called 'formal' properties, *e.g.*, properties of numbers. Its utmost range of application seems to be the various universes of what may be called 'material' properties, *e.g.*, properties of physical things.

Let us assume that we are clear about the Universe(s) of Properties to which the Principle of Determinism is intended to apply. It may now be suggested that the principle means that all properties in any such universe are Determined Properties in regard to the rest of the universe.

This identification of ϕ_0 with the rest of the universe in question—*i.e.*, with the universe exclusive of the Determined Property itself—is, however, open to an objection. For, given a property, we can always define a Sufficient Condition of it in each instance of its occurrence. *E.g.*, the property 'being the n:th positive instance of H when H is being counted in the way R' is a Sufficient Condition of H. Some such condition, moreover, is present in each positive instance of H. Thus H is a Determined Property in its own universe (exclusive of itself).

The trivialities arising from the possibility of defining such 'artificial' properties are avoided if ϕ_0 is identified, for any given property in the universe in question, not with all the remaining properties as such in that universe, but with all the remaining properties of which the given property is logically totally independent.

Thus we may suggest the following definition of a property being 'absolutely' determined:

A property is a Determined Property, if it is a Determined Property in the set of all properties (in its universe) of which it is logically totally independent.

We may then suggest the following interpretation of the principle that 'nothing occurs without a cause' or the Principle of Determinism:

All properties (in a certain universe) are Determined Properties.

(It is fairly obvious that still further restrictions and qualifications are needed in order to make the Principle of Determinism plausible, but they will not be discussed in this inquiry.)

Finally, if a partial analysis of causality in terms of conditions
is accepted, the Logic of Conditions would come to include that
important branch of the logic of actual reasoning, both in the
sciences and in practical life, which may be called the Logic of
Causal Analysis. This would imply that whenever we reason in
causal terms, whether in connexion with scientific observation
and experimentation, or for or against opinions on matters of
ethics and politics, whether in order to trace the interconnexion
of events in the history of nations or in that of single individuals,
we constantly apply the Logic of Conditions (and the principles
of inductive inference based upon it).

Simple though this logic is in theory, investigation would
probably show that there are, in practice, more errors com-
mitted through breaches of its principles than through breaches
of those in any other elementary branch of logical thinking.
Even in philosophic and scientific works there is often confusion
of thought, arising from neglect of the principal distinctions in
regard to the conditioning relation, and their implications. We
shall later have occasion to give some examples of this. They will
serve to confirm the opinion that the Logic of Conditions possesses a not inconsiderable didactic value for training our
minds to reason clearly.

Note.—The first attempt known to me at a systematic treatment of the Logic of
Conditions is to be found in C. D. Broad's article " The Principles of Demonstrative Induction I " in *Mind 39* (1930).

The Logic of Conditions was further developed in my thesis *The Logical Problem
of Induction* (1941), in my paper "Några anmärkningar om nödvändiga och
tillräckliga betingelser " in *Ajatus 11* (1942), and in Broad's article "Hr. Von
Wright on the Logic of Induction I " in *Mind 53* (1944).

3. Statistical Laws

Consider two properties ('factors', 'terms'). One might
suggest that there are two basic ways in which they can be
connected by law. The first asserts an invariable connexion:
all positive instances of the one property are also positive
instances of the other, or, to put it otherwise, the one property

is totally included in the other. The second asserts a statistical correlation: a *proportion p* of the positive instances of the one property are positive instances of the other as well, or, to put it otherwise, a proportion *p* of the one property is included in the other.

In the case of total inclusion the law is a Universal Implication or Equivalence. In the case of partial inclusion we shall call the law a Statistical Law.

The traditional attitude has long been, to regard Statistical Laws as somehow philosophically subordinate to laws of the 'causal type' asserting invariable connexions. This attitude, however, is gradually being reversed, thanks to the development of modern science and, not least, to the rise of Quantum Physics. It seems reasonable to ask whether all uniformities of nature are not at bottom statistical only, and whether Universal Implications and Equivalences are not 'idealizations' or 'approximations' on the basis of high-degree statistical correlations. A Logic of Induction, therefore, which takes account only of laws asserting invariable connexions, is in constant danger of losing touch with the actual procedures of science.

In order to bring out the relevance of the above objection and warning, it will be our first duty to scrutinize the logical foundations of the division of Laws of Nature into invariable and statistical uniformities. It will be seen that the difference between the two types of law is not merely that of a simple contrast, but is more intricate than would appear at first sight.

* * * * *

It is of great importance to stress that the phrase 'a certain proportion of (the property denoted by) H is included in (the property denoted by) A', makes sense only on condition that the property H is either

or
 i. finite
 ii. replaced by a sequence H, R.

Thus the notion of 'a proportion of an infinite population' is, as such, meaningless. It need not, however, be rejected; we can make its use legitimate by introducing a way of counting the population.

We shall soon be in a position to see why the idea of a proportion, if not explicitly restricted to finite populations, must be taken in relation to a way of counting things, *i.e.*, of ordering them into a sequence. The tendency to ignore this relational aspect of proportions is particularly strong in cases where there exists a temporal succession of the things, determining a 'natural' way of counting them.

Let us first consider the alternative that H is finite. In this case H has a cardinal number n. The same is true of the product of H and any property. Thus, in particular, $A\&H$ has a cardinal number m $(m\leqslant n)$. The ratio $m:n$ indicates the proportion of H included in A or, as we may also put it, the relative frequency of A's among H's.

Let us then turn to the alternative in which we have a sequence H, R. For the part of the sequence which consists of the n first positive instances of H (p. 52) we introduce the symbol $_RH_n$. This part can be treated as a finite property. It has a cardinal number (which equals n if R is a dense way of counting H beginning from 1). So also has any part of it which originates from the product of H and any property, *e.g.*, A. The ratio of the cardinal numbers of $A\&_RH_n$ and $_RH_n$ indicates the relative frequency of A's among the n first H's. That this ratio has the value p_n we can express by $Fr(A, H, R, n, p_n)$.

The values p_n form a sequence of rational numbers.

Let lim (p_n, p) be the case (for the definition of lim see p. 53). On this assumption we say that a proportion p of H is included in A, or that the relative frequency of A's among H's is p, when H is counted in the way R. This proposition can be expressed $(n)Fr(A, H, R, n, p_n)\&lim(p_n, p)$. As an abbreviation for the expression we introduce $F(A, H, R, p)$.

The proportion is thus the limiting frequency of A's among the n first H's, n being indefinitely increased. The phrase 'the n first H's' indicates the essential fact that the proportion is viewed relative to a way of counting H.

If H is finite, then the sequence of values p_n terminates at a certain cardinal m. We make the sequence infinite by adopting the convention that, for $n>m$, p_n always equals p_m. This makes the concept of a limiting frequency apply also to finite populations.

If H is infinite, then the Statistical Law $F(A, H, R, p)$ can be

neither verified nor falsified on the basis of statistical observations alone, *i.e.*, on the basis of observed relative frequencies in
finite parts or 'samples' from H, R. This exemption of Statistical
Laws from proof and disproof by experience has been the subject-matter of much discussion. We need not enter into the
problem here. But it ought to be stressed that the above-
mentioned peculiarity of Statistical Laws in respect of decida
bility, does not affect the logical legitimacy (consistency) of the
concept of a proportion as defined in this book.

If p is 0 or 1, we call the proportion extreme. If p is 0, we
call the proportion minimal or imperceptible. If p is greater
than 0, we call the proportion perceptible. If p is 1, we call the
proportion maximal.

The proposition that a maximal proportion of H is included
in A or that 100 per cent of the H's are A's, should not be confused with the proposition that H is totally included in A or
that *all* H's are A. The latter proposition entails the former, but
not *vice versa*. That 100 per cent of the H's are A's is logically
compatible with the existence of even an infinite number of
H's which are not A's.

Analogously, the proposition that a minimal proportion of
H is included in A or that 0 per cent of the H's are A's should
not be confused with the proposition that *no* H's are A's. The
limiting frequency of primes, for instance, among the cardinals
is 0 and yet there is, according to a well-known theorem, an
infinite number of primes.

Bearing this in mind, we shall say that '(all or) practically all'
H's are A, if a maximal proportion of H is included in A, and
that '(no or) practically no' H's are A, if a minimal proportion
of H is included in A.

The necessity of taking the notion of a proportion, when
applied to infinite populations, as relative to a way of counting,
is clearly seen from the fact that by re-ordering a sequence
H, R, *i.e.*, by replacing R with a different way of counting, R',
we may alter or even totally destroy the limiting frequency of
A's among the n first H's. This is best shown by means of an
example.

Let us assume that one dense way R of counting H would be
such that exactly every odd-numbered member of the sequence

was a positive instance of A and every even-numbered not. Let us then use another way R' of counting H, thereby re-ordering the sequence in such a manner that between any two positive instances of A two negative instances of A regularly occur. The original and the re-ordered sequence, *i.e.*, H, R and H, R', contain the same members, *i.e.*, the positive instances of H; but whereas the limiting frequency of A's among the n first positive instances of H is $1 : 2$ in the first sequence, it is $1 : 3$ in the second. It should be added that the re-ordering affects the proportion solely on condition that it is not restricted to a finite part of the original sequence.

In the example just mentioned the limiting frequency was altered as a consequence of re-ordering. But it can also be destroyed. Let the members of the above sequence H, R be re-ordered by means of a new way R' of counting, such that the first two members of H, R' are the first two members of H, R which are not A, the next four members of H, R' are the first four members of H, R which are A, the next six members of H, R' are the last six of the first eight members of H, R which are not A, the next twelve members of H, R' are the last twelve of the first sixteen members of H, R which are A, the next twenty-four members of H, R' are the last twenty-four of the thirty-two first members of H, R which are not A, and so on. (The rule should now be obvious to the reader.) In this sequence the relative frequency of A's among the n first H's perpetually oscillates between the two extremes $1 : 3$ and $2 : 3$ inclusive. We recognize (cf. p. 53) that these extremes, and also any value between them, are partial limits of p_n, or of the relative frequency of A's among the n first H's.

The partial limits can be altered by re-ordering, but, in virtue of the Bolzano-Weierstrass Principle (p. 53), cannot be destroyed.

We have now completed our analysis of the notion of partial inclusion or of Statistical Law. It remains to consider what its relation is to total inclusion or Universal Implication.

What it means for a sequence of real numbers to approach a limit was defined in expression (1) of Chap. II, §5. It was, however, observed, that if the numbers fall within a finite interval, the approach to a limit can also be defined by means of expression (4) of Chap. II, §5. Since the value of a relative frequency must lie in the interval from 0 to 1 inclusive, it follows that both definitions apply to Statistical Laws.

The proposition expressed in (1) of Chap. II, §5 can be viewed as a Universal Implication to the effect that every value δ which has the 'property' of being greater than 0, also has the further 'property' of being associated with a point of convergence m of a certain sequence towards a value p. An instance of this Universal Implication is provided by every proposition to the effect that there exists such a point of convergence for a particular value δ. The instances of the Universal Proposition in question are thus Existential Propositions.

The proposition expressed in (4) of Chap. II, §5 can be viewed as a Universal Implication to the effect that every value q which has the 'property' of being different from p, also has the further 'property' of being associated with a point of divergence m of a certain sequence from the value q itself. An instance of this Universal Implication is provided by every proposition to the effect that there exists such a point of divergence for a particular value q. The instances of the Universal Proposition in question are thus Existential Propositions.

Thus it is possible to view the Statistical Law expressed in $F(A, H, R, p)$ as a Universal Implication. This, moreover, can be done in two different ways. Depending upon which way is chosen, the instances of the law are propositions to the effect that, in the sequence of relative frequencies p_n of A's among the n first H's, there exist either points of convergence towards a given value p associated with particular values of a quantity δ which is greater than 0, or points of divergence from particular values of a quantity q which is different from p.

$$* \quad * \quad * \quad * \quad *$$

We can now return to the points made at the beginning of this section.

It has been shown that our original definition of Laws of Nature as Universal Implications or Equivalences is, in fact, sufficiently general to embrace also Statistical Laws. This means that inductive inference leading to Statistical Laws is a sub-species of inductive inference of the second order, as previously defined by us. Thus the logical study of induction, as undertaken in this book, is relevant not only to a narrow type of inductive conclusion, but is the study of the most general and comprehensive pattern of such inference.

All this, however, does not at all stultify the demand for a special study of inductive inference as leading to Statistical Laws. It is to be regretted that not much attention is paid to this study within the limits of the present book. The subject is comparatively new and still remains largely a *desideratum* in Inductive Logic.

Just as we can distinguish between induction of the first order leading to predictions, and induction of the second order leading to theories (laws), so we can also distinguish between induction leading to statistical predictions and induction leading to statistical theories (laws). The part of the Logic of Induction which studies the making of statistical predictions and theories (laws) may be called the Logic of Statistical Inference.

This part of the Logic of Induction is, however, not the only one deserving a particular study.—Another fundamental division of the Laws of Nature, beside that into propositions dealing with total and partial inclusion, is the division into Qualitative and Quantitative Laws, the latter being those in which the properties ('terms', 'factors') related by law are, in one sense or other, measurable. Here again we could regard Qualitative Laws as the all-embracing type of law and show that Quantitative Laws are a sub-species.

A systematic treatment of what might be called the Logic of Quantitative Induction is another urgent *desideratum* in Inductive Logic. It is astonishing how little has actually been done in this important field of research. Bacon's description of the *Tabula graduum sive comparative* and Mill's account of the Method of Concomitant Variations are almost all that is available, and these contributions, it seems, do not really contain anything that is peculiar to Quantitative Laws of Nature, but merely apply to such laws results emerging from the general Logic of Induction as treated by the authors mentioned.

Chapter Four

INDUCTION AND ELIMINATION

1. The Methods of Induction

IN this and the next two chapters, the relation between premisses and conclusions of inductive arguments of the second order will be studied, considerations as to probability being for the present excluded.

The premisses of an inductive argument can be characterized as non-demonstrative evidence in favour of the conclusion. It is the aim of the various so-called Methods of Induction to afford principles guiding us in the accumulation and use of this evidence.

It is customary to distinguish between two chief Methods of Induction.

The first method consists simply in the multiplication of things which afford confirmations of a certain law. Or else it consists in the multiplication of premisses in support of the conclusion. For such a method the *number* of things (premisses), as such, is relevant. We call it Induction by Simple Enumeration (*inductio per enumerationem simplicem*) or Enumerative Induction.

It is obvious that from the mere use of Enumerative Induction no (demonstrative) conclusions can be drawn as regards the truth-values of laws.

Whether from such a use of Enumerative Induction conclusions can be drawn as to the probability of laws, has been a matter of controversy. It is a well-known fact that an increase in the number of confirming instances of a law is regarded, as a rule, as contributing to its probability. It is uncertain, however, whether this increase in probability can be regarded as a genuine effect of the increasing number of premisses in support of the conclusion, or whether it must be attributed to the hidden force of some other inductive method operating 'behind the scenes.' This question will be a chief object of discussion in Chap. IX.

The second method consists in the examination of things affording confirmations of a certain law, with regard to their resemblance and difference. It may add new premisses to already given ones, but the *number* of premisses is not in itself relevant. We shall presently introduce a name for this method also.

Use of this second method does permit (demonstrative) conclusions as to the truth-values of laws. Remembering what was said previously (p. 44 f.) about verifying and falsifying instances of Universal Propositions, it is clear that these conclusions are always in the negative, *i.e.* relate to the falsehood of laws. The method, therefore, has rightly been called 'a negative approach to the truth.' What is meant by this can be made clearer as follows:

Consider a set of Universal Implications or Equivalences. It will be assumed that the properties in question are all from the same Universe of Properties.

Any one of the members of the corresponding Universe of Things either affords a confirming or a disconfirming instance of any one of the given laws, *i.e.*, the Universal Implications or Equivalences.

Consider next a set of things from this universe. It can be used for dividing the above set of Universal Implications or Equivalences into two groups. The first group consists of those laws, of which *all* the members of the set of things afford confirmations. The second group consists of those of which at least one member of the set of things affords a disconfirming instance, or else of which *not all* the members of the set afford confirmations. Considering the asymmetry of Universal Propositions in respect of verification and falsification (p. 44 f.), we call the laws of the first group compatible and those of the second group incompatible with the set of things.

Let there now be two sets of things. Let the first set be included in the second, *i.e.*, let any member of the first set be a member of the second set also. It is clear that the group of laws which are compatible with the second set of things is included in the group of laws which are compatible with the first set of things. For any member of the former group of laws *must* be compatible with both sets of things, whereas any member of the latter group of laws *may* be incompatible with the second set of things.

Consequently, an increase in the number of things affording confirmations, or else in the number of premisses, of a certain inductive conclusion, may effect a decrease in the number of laws which are compatible with the set of things (premisses). Whether and in what degree such a decrease will take place depends entirely upon the resemblance and difference between the things. To account for the nature of this dependence has traditionally been one of the chief tasks of the Logic of Induction.

Induction which is not by simple enumeration is thus a procedure of elimination or exclusion of laws from compatibility with facts. We shall call this method Induction by Elimination or Exclusion, or Eliminative Induction. It is a negative approach to the truth in the sense that it makes the conclusion of an inductive argument appear to stand out more and more, so to speak, among a number of initially 'concurrent' conclusions (laws). Its logical mechanism rests on the fundamental, though trivial, fact that no confirming instance of a law is a verifying instance, but that any disconfirming instance is a falsifying instance. It is the immortal merit of Francis Bacon to have first realized the importance of this fact for the logical study of induction.

Induction by Elimination evidently represents a much more advanced mode of inductive procedure than Induction by 'simple' Enumeration. It has been claimed that Eliminative Induction is the genuinely 'scientific' and 'methodical' way of inductive reasoning, as against the 'unscientific' and 'un-methodical' procedure of Enumerative Induction. To what extent and in what sense this claim can be justified will be investigated in Chap. IX.

2. The Method of Elimination. General Remarks

A convenient mode of studying the logic of elimination will be to study the eliminative method in its application to certain basic questions concerning nomic connexions between properties. Those questions are:

i. What are the Necessary Conditions of a given property?
iiα. What are the Sufficient Conditions of a given property?
iiβ. What are the Sufficient Conditions of a given property in a given positive instance of it?
iii. What are the Necessary-and-Sufficient Conditions of a given property?

The division of the second question, but not of the first, into two sub-questions, should be understood against the background of the important difference between Sufficient and Necessary Conditions which was pointed out in *T24* and *T25* of Chap. III, §2. Whereas, in any positive instance of a property, every one of its Necessary Conditions *must* be present, any given one of its Sufficient Conditions *may* be absent.

The given property we call the conditioned property.

The properties which we are seeking we call (actual) conditioning properties. To any one of the (actual) conditioning properties there corresponds what we shall call an actual nomic connexion.

The properties among which the (actual) conditioning properties are sought, we call initially possible conditioning properties. To any one of the initially possible conditioning properties there answers what we shall call an initially possible nomic connexion.

By a datum of elimination, we shall understand any true proposition to the effect that a certain one of the initially possible conditioning properties has been excluded.

The data of elimination are afforded by things which are instances of the conditioned property (and of the actual and possible conditioning properties). In the case of question i, above, the data are afforded by positive instances of the conditioned property which are negative instances of some of the initially possible conditioning properties. In the case of question ii

(α and β), the data are afforded by negative instances of the conditioned property which are positive instances of some of the initially possible conditioning properties. In the case of question iii, the data are afforded either by positive instances of the conditioned property which are negative instances of some of the initially possible conditioning properties, or by negative instances of the conditioned property which are positive instances of some of the initially possible conditioning properties.

Supposing the instances of the conditioned property to be denumerable (p. 51), we can order them into a sequence $x_1, \ldots, x_n, \ldots$

Let ϕ_0 denote a set of initially possible conditioning properties of a given conditioned property.

According to what was said in the preceding section of this chapter, we can divide the initially possible nomic connexions which answer to the members of ϕ_0 into two groups, according to whether or not they are compatible with x_1, i.e., according to whether x_1 affords a confirming or a disconfirming instance of the respective nomic connexions (laws). The set of possible conditioning properties which answer to the members of the first of these groups of laws we denote ϕ_1. The members of ϕ_1 are called the remaining possible conditioning properties of the given conditioned property relative to x_1. ϕ_1 is co-extensive with ϕ_0 or not, according to whether x_1 affords data of elimination or not.

Similarly, we can divide the initially possible nomic connexions which answer to the members of ϕ_0 into two groups, according to whether or not they are compatible with x_1 *and* x_2. On the basis of this division we introduce the set of properties ϕ_2, the members of which are called the remaining possible conditioning properties of the given conditioned property relative to x_1 and x_2. ϕ_2 is included in ϕ_1. ϕ_2 is co-extensive with ϕ_1 or not, according to whether or not x_2 affords data of elimination, not already afforded by x_1.

Analogously, we introduce ϕ_3, *etc.*

Thus we produce a sequence $\phi_1, \ldots, \phi_n, \ldots$

Finally, we divide the initially possible nomic connexions which answer to the members of ϕ_0 into two groups, according to whether or not they are compatible with all things in the universe, *i.e.*, according to whether or not they are true propositions. The set of possible conditioning properties which answers to the members of the first of these groups of laws we denote ϕ. The members of ϕ are the actual conditioning properties of the given conditioned property, among the set ϕ_0 of initially possible conditioning properties.

It can be shown that the sequence of properties $\phi_1, \ldots, \phi_n, \ldots$ approaches as its limit the property ϕ. For, first, every property ϕ_{n+1} is included in ϕ_n. Secondly, whatever is a positive instance of ϕ, *i.e.*, every actual conditioning property (in ϕ_0) of the given conditioned property, is also a positive instance of every one of the properties $\phi_1, \ldots, \phi_n, \ldots$, *i.e.*, of the remaining possible conditioning properties (in ϕ_0). And thirdly, whatever is a positive instance of all the properties $\phi_1, \ldots, \phi_n, \ldots$, *i.e.*, belongs to all the fields of remaining possible conditioning properties, is also a positive instance of ϕ, *i.e.*, is an actual conditioning property of the given conditioned property.

(It should be carefully observed that the last statement only holds good on condition that the sequence $x_1, \ldots, x_n, \ldots$, on the basis of which the sequence $\phi_1, \ldots, \phi_n, \ldots$ has been defined, is a denumeration either of *all* instances of the conditioned property, *i.e.*, of *all* things in the universe in question, or at least of all instances of the conditioned property which may afford a datum of elimination, *i.e.*, of at least all positive or all negative instances of the conditioned property, as the case may be (cf. above p. 87 f.). We shall have occasion to discuss the relevance of this condition later, in Chap. IX.)

Thus we have shown that the three conditions are fulfilled (cf. above p. 54 f.) which define $lim(\phi_n, \phi)$. The range of actual conditioning properties of a given conditioned property is the limiting extension approached by the ranges of remaining possible conditioning properties. One might also say that the range of actual conditioning properties marks the limit of elimination.

The four questions stated at the beginning of this section correspond to four basic modes of application of the general Method of Elimination. For these modes or sub-methods we shall use separate names, *viz.* the Direct Method of Agreement, corresponding to question i, the Inverse Method of Agreement, corresponding to question iia, the Method of Difference, corresponding to question iiβ, and the Joint Method, corresponding to question iii.

Of each of the four sub-methods of the Method of Elimination we may further distinguish two cases. We shall call them the Simple Case and the Complex Case respectively. They differ in the composition of the set of initially possible conditioning properties of a given conditioned property.

Let ϕ_0 again denote a set of logically totally independent properties from one and the same universe.

In the Simple Case, the set of initially possible conditioning properties consists of:

i.   the members of some such set ϕ_0 and the negation-properties of the members, or:

ii.   those of the properties mentioned in i, which are present in a given positive instance of the conditioned property (Method of Difference).

In the Complex Case, the set of initially possible conditioning properties consists of:

i.   the members of some such set ϕ_0 and all sums of some m $(0 \leqslant m \leqslant n)$ of some n $(1 \leqslant n)$ members of ϕ_0 and the negation-properties of the remaining $n-m$ members (Direct Method of Agreement), or:

ii.   the members of some such field ϕ_0 and all products of some m $(0 \leqslant m \leqslant n)$ of some n $(1 \leqslant n)$ members of ϕ_0 and the negation-properties of the remaining $n-m$ members (Inverse Method of Agreement), or:

iii. those of the properties mentioned in ii which are present in a given positive instance of the conditioned property (Method of Difference), or:

iv. the members of some such set ϕ_0 and all presence-functions—the tautology and the contradiction, however, being excluded—of some n $(1 \leqslant n)$ members of ϕ_0 (Joint Method).

The Simple Case of each of the sub-methods is included in the Complex Case of the same sub-method as an extreme alternative. For expository as well as historical reasons it is, however, convenient to treat the two cases separately.

Let there be a set of initially possible conditioning properties in the sense of the Simple Case (i or ii). Suppose that k only of the corresponding initially possible nomic connexions are compatible with a given set of things. (Cf. p. 85.) Then k is said to determine or to measure the state of analogy among the things, in respect of the initially possible conditioning properties (or nomic connexions). If k equals 1, we speak of Perfect Analogy. If k equals 0, we speak of Total Elimination.

Let there be a set of initially possible conditioning properties in the sense of the Complex Case i ,ii, or iii. Consider the sub-set which consists only of n-termed products or sums. Suppose that k only of the corresponding initially possible nomic connexions are compatible with a given set of things. Then k is said to determine or to measure the state of analogy on the n-level among the things, in respect of the initially possible conditioning properties (or nomic connexions). If k equals 1, we speak of Perfect Analogy on the n-level. If k equals 0, we speak of Total Elimination on the n-level.

Let there be a set of initially possible conditioning properties in the sense of the Complex Case iv. Consider the sub-set which consists of properties which are presence-functions of some n properties in ϕ_0 but not of a lesser number of properties. Suppose that k only of the corresponding initially possible nomic connexions are compatible with a given set of things. Then k is said to determine or to measure the state of analogy on the n-level among the things, in respect of the initially possible conditioning properties (or nomic connexions). If k equals 1, we speak of Perfect Analogy on the n-level. If k equals 0, we speak of Total Elimination on the n-level.

By Absolutely Total Elimination we understand Total Elimination on all levels. By Absolutely Perfect Analogy we understand Perfect Analogy on a certain level in combination with Total Elimination on all other levels.

The case of (Absolutely) Perfect Analogy and of (Absolutely) Total Elimination are of particular interest, not least from the point of view of scientific practice. We shall therefore devote special attention to them in the logical examination of the inductive methods.

*   *   *   *   *

The traditional Logic of Induction has, with a few non-systematic exceptions, been limited to a treatment of what is here called the Simple Case of the sub-methods of the general Method of Elimination. This 'traditional' treatment is substantially identical with the contributions of Bacon and Mill to the subject. (See Chap. VI, §4.)

It is, however, clear that a treatment which claims to provide an adequate logical instrument for reconstructing the actual procedures of science, must account for the Complex Case also. The importance of solving the various problems of formal logic which arise in connexion with this account (see §§ 6–8 of this chapter) should not be overrated. But it seems to me undeniable that the study of a Logic of Induction which advances beyond the tables of Bacon and the canons of Mill is of some interest, not only to the professional logician and philosopher, but also to anyone who is concerned with the design of experiments and the methodical performance of observations in scientific research.

The demand for a particular treatment of the Complex Case is also made urgent by the fact that there is a tendency, in the development of science, to emphasize more and more the complexity of conditional relations and nomic connexions. Thus, to mention only one example, the rise of *Gestalt*-Psychology and the introduction of laws of a new type, called *Gestalt*-laws, (which are sometimes claimed to be radically different from the laws of classical physics and traditional associationist psychology), would seem to be essentially a transition from nomic connexions in the sense of the Simple Case of the inductive methods, to nomic connexions in the sense of the Complex Case, in the realm of biological and mental phenomena.

It is not unlikely that an analysis of the notion of *Gestalt* and kindred ideas in terms of the Logic of Conditions would contribute to the clarification of a number of obscure questions in the philosophic foundations of modern science. The *Gestalt*, in some uses of the word, seems to be a disjunction of Substitutable Requirements within a 'frame' of (Indispensable) Contributory Conditions of a phenomenon.

In view of all this, the extension of the 'classical' Logic of Induction treating only the Simple Case, to a 'modern' Logic of Induction dealing with the Complex Case also, might be regarded as a counterpart, in the study of scientific method, to a main trend in the development of scientific ideas.

3. The Method of Agreement. The Simple Case

A. The Direct Method.

The conditioned property is called H.

The (actual) conditioning properties are Necessary Conditions of H.

Let ϕ_0 denote a set of logically totally independent properties from the same universe as H. It is understood that H is logically totally independent of the properties in ϕ_0.

The initially possible conditioning properties are the properties in ϕ_0 and their negation-properties.

The method is based on the following Principle of Elimination: Whatever is absent in the presence of the conditioned property cannot be a Necessary Condition of it.

Consequently, the data of elimination are afforded by positive instances of the conditioned property which are negative instances of some of the initially possible conditioning properties. (Cf. above p. 87.)

The remaining possible conditioning properties, relative to n positive instances of the conditioned property, are those of the initially possible conditioning properties which are common to all (present in every one of) the n instances.

Given a positive instance of H, any one of the properties in ϕ_0 is either present or absent in this instance. If a property is present, its negation-property is absent, and if a property is absent, its negation-property is present. It follows that any positive instance of H excludes or eliminates exactly half of the initially possible conditioning properties from being Necessary Conditions of H.

We shall call the effect of elimination exerted separately by each thing which may afford a datum of elimination, the immediate eliminative effect.

Two things are said to differ (or to vary) in respect of a property, if one of the things is a positive and the other thing a negative instance of the property. Two things are said not to differ in respect of a property, if the things are either both positive or both negative instances of the property.

Let there be a positive instance of H. Then any other positive instance of H eliminates as many initially possible conditioning properties, not already eliminated by the first instance, as there are properties in ϕ_0 in respect of which the two instances differ. Generally speaking, any additional positive instance of H eliminates as many initially possible conditioning properties, not already eliminated by previous positive instances of H, as there are properties in ϕ_0 in respect of which the additional positive instance of H differs from *all* the previous positive instances of H.

We shall call the effect of elimination exerted by an additional thing which may afford a datum of elimination, (as opposed to previous things), the additional eliminative effect.

Thus in the Simple Case of the Direct Method of Agreement, the immediate eliminative effect has a constant magnitude equal to half of the initially possible conditioning properties, whereas the additional eliminative effect is proportionate to the (extension of the) sub-set of properties in ϕ_0, in respect of which an additional positive instance of the conditioned property differs from every one of a number of previous positive instances of the conditioned property.

If two positive instances of H differ in respect of all the properties in ϕ_0, we have attained Total Elimination of the initially possible conditioning properties. If two positive instances of H differ in respect of all but one of the properties in ϕ_0, we have attained a Perfect Analogy.

Thus in the Simple Case of the Direct Method of Agreement Total Elimination and Perfect Analogy respectively are attainable in a minimum of two 'steps' ('observations', 'experiments').

B. The Inverse Method.

The conditioned property is called H.

The (actual) conditioning properties are Sufficient Conditions of H.

Let ϕ_0 mean the same as in the Direct Method.

The initially possible conditioning properties are the same as in the Direct Method.

The method is based on the following Principle of Elimination: Whatever is present in the absence of the conditioned property cannot be a Sufficient Condition of it.

Consequently, the data of elimination are afforded by negative instances of the conditioned property which are positive instances of some of the initially possible conditioning properties. (Cf. above p. 87.)

The remaining possible conditioning properties, relative to n negative instances of the conditioned property, are those of the initially possible conditioning properties which are absent in every one of the n instances.

For analogous reasons to those in the Direct Method, any negative instance of H eliminates exactly half of the initially possible conditioning properties from being a Sufficient Condition of H.

Similarly, any additional negative instance of H eliminates as many initially possible conditioning properties, not already eliminated by previous negative instances of H, as there are properties in ϕ_0 in respect of which the additional instance differs from all the previous instances.

Thus in the Simple Case of the Inverse Method of Agreement the immediate and the additional eliminative effects have the same magnitude as in the Simple Case of the Direct Method of Agreement.

If two negative instances of H differ in respect of all the properties in ϕ_0, we have attained Total Elimination, and if they differ in respect of all but one of the properties in ϕ_0, we have attained Perfect Analogy. Thus also in the Simple Case of the Inverse Method of Agreement, these 'ideal' states of elimination are attainable in a minimum of two 'steps' ('observations,' 'experiments').

*   *   *   *   *

The Method of Agreement, in its Direct and Inverse form, is the reasoning mechanism underlying inductive (or 'causal') arguments from a minimum of constancy amid a maximum of variation. It brings out the relevance of the methodological device, well-known from scientific observation and experimentation alike, of 'varying the circumstances.'

4. The Method of Difference. The Simple Case

The conditioned property is called H.

The (actual) conditioning properties are Sufficient Conditions of H.

Let ϕ_0 mean the same as in the Method of Agreement.

The initially possible conditioning properties are those of the properties in ϕ_0 and their negation-properties which are present in a given positive instance of H.

The Principle of Elimination is the same as in the Inverse Method of Agreement.

Consequently, the remaining possible conditioning properties (relative to the given positive and n negative instances of the conditioned property) are defined in the same way as in the Inverse Method of Agreement.

In Contrast to the Simple Case of the Method of Agreement, the immediate eliminative effect exerted by any thing which may afford a datum of elimination does not have a constant magnitude in the Simple Case of the Method of Difference. If a negative instance of H has no initially possible conditioning properties in common with the given positive instance of H, then it is wholly ineffective for the purpose of elimination. If, again, it has all initially possible conditioning properties in common with the given positive instance of H, it effects Total Elimination, and if it has all but one of the initially possible conditioning properties in common with the positive instance, it establishes a Perfect Analogy.

Let there be a negative instance of H. Then any other negative instance of H eliminates as many initially possible conditioning properties as there are properties in ϕ_0 in respect of which it agrees with the given positive instance of H and differs from the first negative instance of H. Generally speaking, any additional negative instance of H eliminates as many initially possible conditioning properties as there are properties in ϕ_0 in respect of which it agrees with the given positive instance of H and differs from all the previous negative instances of H.

Particular importance has traditionally been attached to the case where one negative instance of the conditioned property suffices to establish a Perfect Analogy. This is the unique case which Mill describes under the title Method of Difference.

*　　*　　*　　*　　*

The method of Difference is the reasoning mechanism under-lying inductive (or 'causal') arguments from a minimum of variation amid a maximum of constancy. It brings out the relevance of the methodological device, well-known in particu-lar from scientific experimentation, of 'removing a factor which leaves other circumstances unchanged.'

Note.—Suppose that one negative instance of H is compared with positive instances of H. Then any property which is common to the negative instance and at least one of the positive instances is excluded from being a Sufficient Condition of $\bar{H}$. It follows by contraposition that the negation-property of any such property is excluded from being a Necessary Condition of H.

This fact might suggest that there are two forms of the Method of Difference, one in which the (actual) conditioning properties are Sufficient and another in which they are Necessary Conditions of the conditioned property, exactly as there are two forms, the Direct and the Inverse, of the Method of Agreement. This suggestion, however, is misleading. For the negation of any property which is present in at least one of the positive instances of H cannot be a Necessary Con-dition of H, quite independently of whether or not it is present in the negative instance of H. Thus the result to which the suggested 'inverse' application of the Method of Difference would lead, can already be anticipated from the negative instances of H according to the Direct Method of Agreement. The peculiar interest of the Method of Difference arises solely from the fact that a property's Sufficient Conditions, as opposed to its Necessary Conditions, need not all be present in any positive instance of the property. (Cf. above p. 72.)

The above was not clearly recognized by me in the paper *Några anmärkningar om nödvändiga och tillräckliga betingelser*, where I distinguished between two forms of the Method of Difference in analogy with the two forms of the Method of Agreement.

5. *The Joint Method. The Simple Case*

The conditioned property is called H.

The (actual) conditioning properties are Necessary-and-Sufficient Conditions of H.

Let ϕ_0 mean the same as in the Method of Agreement.

The initially possible conditioning properties are the same as in the Method of Agreement.

The method is based on the following Principle of Elimination: Whatever is absent in the presence or present in the absence of the conditioned property, cannot be a Necessary-and-Sufficient Condition of it.

Consequently, the data of elimination are afforded by positive instances of the conditioned property which are negative instances of some of the initially possible conditioning properties and by negative instances of the conditioned property which are positive instances of some of the initially possible conditioning properties. (Cf. above p. 88.)

The remaining possible conditioning properties, relative to m positive and n negative instances of the conditioned property, are those of the initially possible conditioning properties which are present in every one of the m positive instances and absent in every one of the n negative instances.

As will be seen, the Principle of Elimination which is used in the Joint Method is the disjunction of the eliminative principles used in the Direct and Inverse Method of Agreement, and in the Direct Method of Agreement and the Method of Difference respectively. It follows that each one of the methods mentioned can be used separately for ascertaining Necessary-and-Sufficient Conditions. The remaining possible Necessary Conditions, relative to n positive instances of a conditioned property, are the same as the remaining possible Necessary-and-Sufficient Conditions, relative to the same instances of the conditioned property. Similarly, the remaining possible Sufficient Conditions, relative to n negative or to one positive and n negative instances of a conditioned property, are the same as the remaining possible Necessary-and-Sufficient Conditions, relative to the same instances of the conditioned property.

It is, however, also possible to make a joint use of the methods for ascertaining Necessary-and-Sufficient Conditions. The idea is to compare a set of remaining possible Necessary Conditions and a set of remaining possible Sufficient Conditions of the conditioned property. The common members of both sets will constitute a set of remaining possible Necessary-and-Sufficient Conditions of the property.

We obtain the remaining possible Necessary Conditions after resort to the Direct Method of Agreement. We obtain the remaining possible Sufficient Conditions either after resort to the Inverse Method of Agreement or to the Method of Difference.

Thus there are two possibilities of a joint use of the methods. The first is to combine the Direct Method of Agreement with the Inverse Method of Agreement. The second is to combine the Direct Method of Agreement with the Method of Difference. We shall call the method originating from the first combination the Double Method of Agreement, and the method originating from the second the Joint Method of Agreement and Difference.

The fact that there are two forms of the Joint Method has introduced considerable confusion into traditional descriptions of the method.

A. The Double Method of Agreement.

Let there be *m* positive instances of *H* and *n* negative instances of *H*.

The effect of elimination depends here upon two factors, *viz.*, variation among the members of the two sets of instances taken separately, and resemblance between the members of the two sets taken jointly. Any property which is not common to *all* the positive instances, and the negation of any property which is not common to *all* the negative instances, cannot be a Necessary-and-Sufficient Condition of *H*. On the other hand, any property which is common to at least one positive and one negative instance cannot be a Necessary-and-Sufficient Condition of *H*.

We shall call the effect of elimination which depends upon the first factor the variation-effect, and that which depends upon the second factor the resemblance-effect.

The resemblance-effect is independent of the variation-effect. This means that the effect of elimination due to resemblance between the members of the two sets of instances taken jointly may be considerable, though there is little or even no variation among the members of the two sets taken separately. Alternatively, the remaining possible Necessary Conditions, relative to the *m* positive instances of *H*, may be many, and so may the remaining possible Sufficient Conditions, relative to the *n* negative instances of *H*. But the common members of the two sets of remaining possible conditions, *i.e.*, the remaining possible Necessary-and-Sufficient Conditions, may nevertheless be few.

The variation-effect is not, however, independent of the resemblance-effect. That there is variation in respect of a certain property among the members of a set of things means (cf. p. 94.) that the property is present in at least one member and absent in at least one member of the set. It follows that if there is variation in respect of a property, both among a set of positive, and among a set of negative instances of H, then there is resemblance in respect of the property and in respect of its negation-property between the members of the two sets of things taken jointly. For then there exists at least one positive and at least one negative instance of H which agree in respect of the property in question, and also at least one positive and at least one negative instance of H which agree in respect of its negation-property. It is therefore not possible to imagine a situation where there is little or no resemblance between the members of the two sets of instances taken jointly, but much variation among the members of the two sets taken separately.

In view of the above, it is clear that the resemblance-effect is, in fact, responsible for the total effect of elimination, and that the variation-effect can consequently be neglected.—Let there be one positive and one negative instance of H. If the two instances differ in respect of no one of the properties in ϕ_0, then we have attained Total Elimination of the initially possible conditioning properties. If the two instances differ in respect of one and one only of the properties in ϕ_0, then we have attained a Perfect Analogy. Thus Total Elimination and Perfect Analogy are attainable in a minimum number of two 'steps' ('observations,' 'experiments').

B. The Joint Method of Agreement and Difference.

Let there again be m positive and n negative instances of H.

The remaining possible Necessary Conditions are those properties in ϕ_0, or their negation-properties, which are present in all the m positive instances of H. The remaining possible Sufficient Conditions are those properties in ϕ_0, or their negation-properties, which are present in a given one of the m positive instances of H, and absent in all the n negative instances of H. The common members of these two sets of remaining possible conditions constitute a set of remaining possible Necessary-and-Sufficient Conditions of H.

The effect of elimination can again be attributed to two factors, resemblance and variation. The former is in fact responsible for the total effect of elimination. Total Elimination and Perfect Analogy are attainable in a minimum of two 'steps' ('observations', 'experiments').

It is clear that the set of remaining possible Sufficient Conditions will, as a rule, consist of different properties according to the choice of the one positive instance of H. The common members of the sets of remaining possible Necessary Conditions and remaining possible Sufficient Conditions—*i.e.*, the members of the set of remaining possible Necessary-and-Sufficient Conditions—are, however, the same independently of this choice. This is proved as follows:

Suppose that there existed a property A which is one of the initially possible Necessary-and-Sufficient Conditions of H and which is:

i.   common to the sets of remaining possible Necessary Conditions and remaining possible Sufficient Conditions of H, when x is the positive instance of H which has been chosen, and:

ii. not common to the sets of remaining possible Necessary Conditions and remaining possible Sufficient Conditions of H, when y is the positive instance of H which has been chosen.

Part i of the supposition implies that A is present in all the m positive instances and absent in all the n negative instances of H. Part ii of the supposition again implies that A is either absent in y or present in at least one of the n negative instances of H. The two parts of the supposition thus lead to contradictory consequences. It follows by contraposition that the result to which use of the Joint Method of Agreement and Difference will lead us, is independent of the choice of any particular instance of the m positive instances of H to be compared with all the n negative instances of H.

*   *   *   *   *

It can be proved that, given m positive and n negative instances of the conditioned property, the use of the Double Method of Agreement, and of the Joint Method of Agreement and Difference will lead to the same remaining possible Necessary-and-Sufficient Conditions.

Suppose that there existed a property A which is one of the initially possible Necessary-and-Sufficient Conditions of H and which is:

i. one of the remaining possible Necessary-and-Sufficient Conditions, relative to m positive and n negative instances of H, when the Double Method of Agreement has been used, and:

ii. not one of the remaining possible Necessary-and-Sufficient Conditions, relative to the same m positive and n negative instances of H, when the Joint Method of Agreement and Difference has been used.

Part i of the supposition implies that A is present in all the m positive and absent in all the n negative instances of H. Part ii of the supposition again implies that A is either absent in at least one of the m positive instances or present in at least one of the n negative instances of H. Since the two parts of the supposition thus lead to contradictory consequences, it follows by contraposition that the two forms of the Joint Method must lead to concordant results when applied to the same 'material of observation or experimentation.'

6. The Method of Agreement. The Complex Case

A. The Direct Method.

The conditioned property is called H.

The (actual) conditioning properties are Necessary Conditions of H.

Let ϕ_0 mean the same as in the Simple Case.

The initially possible conditioning properties are all properties which are the sums of some m ($0 \leqslant m \leqslant n$) of some n ($1 \leqslant n$) properties from ϕ_0, and the negation-properties of the remaining n-m properties.

INDUCTION AND ELIMINATION

Let us suppose that ϕ_0 has k members. It will be of some interest to calculate, on this supposition, the number of initially possible conditioning properties of different degrees of complexity from 1 to k and also the sum total of initially possible conditioning properties.

The number of sums of (n of) some n properties from ϕ_0 is $\binom{k}{n}$, which can also be written $\binom{k}{n}\binom{k-n}{0}$. The number of sums of n-1 of some n properties from ϕ_0 and the negation-property of the sole remaining property is $\binom{k}{n-1}\binom{k-n+1}{1}$. The number of sums of m of some n properties from ϕ_0 and the negation-properties of the remaining n-m properties is $\binom{k}{n-m}\binom{k-n+m}{m}$. Finally, the number of sums of (0 of some n properties and) the negation-properties of (the remaining) n properties from ϕ_0 is $\binom{k}{n}$, which can also be written $\binom{k}{0}\binom{k}{n}$.

Thus the number of initially possible conditioning properties of degree of complexity n equals $\overset{n}{\underset{m=0}{\Sigma}} \binom{k}{n-m}\binom{k-n+m}{m}$ which equals $\overset{n}{\underset{m=0}{\Sigma}} \binom{k}{n}\binom{n}{m}$. Since $\overset{n}{\underset{m=0}{\Sigma}} \binom{n}{m}$ equals 2^n, we can further simplify the expression to $2^n\binom{k}{n}$.

Consequently, the sum total of initially possible conditioning properties is $\overset{k}{\underset{n=1}{\Sigma}} 2^n\binom{k}{n}$.

When n is 1, we get the number of initially possible conditioning properties of the Simple Case which is $2k$.

When n is 2, we get the number of initially possible conditioning properties of the second degree of complexity. This number is $2k(k$-$1)$.

When n is 3, we get the number of initially possible conditioning properties of the third degree of complexity. This number is $(4k(k$-$1)(k$-$2)) : 3$.

Finally, when n is k we get the number of initially possible conditioning properties of the maximal degree of complexity. This number is 2^k.

The Principle of Elimination is the same as in the Simple Case.

Consequently, the data of elimination are afforded by positive instances of the conditioned property which are negative instances of some of the initially possible conditioning properties.

The remaining possible conditioning properties, relative to n positive instances of the conditioned property, are defined as in the Simple Case.

Suppose again that ϕ_0 has k members. Given a positive instance of H, any one of the k properties in ϕ_0 is either present or absent in this instance. If a property is absent, its negation-property is present. Thus in any positive instance of H, n of the k properties in ϕ_0 are present, and the negation-properties of the remaining k-n properties. The number of properties which are products of some m $(1 \leqslant m \leqslant k)$ of those k properties of the positive instance of H is $\sum\limits_{m=1}^{k} \binom{k}{m}$, which equals 2^k-1. Any one of these 2^k-1 products is present in the positive instance of H. The negation of any one of them is absent. But the negation of a product of some m properties is, according to the Rules of de Morgan, the sum of the negations of the m properties. Thus it is one of the initially possible conditioning properties of H. Since it is absent in the presence of H it is eliminated from being a Necessary Condition of H.

Thus the immediate eliminative effect exerted by any single positive instance of the conditioned property, amounts to $\sum\limits_{m=1}^{k} \binom{k}{m}$ or 2^k-1 initially possible conditioning properties, where k is the number of members of ϕ_0.

When m is 1, we get the immediate eliminative effect of the Simple Case, which is k or half of the initially possible conditioning properties of the Simple Case.

When m is 2, we get the immediate eliminative effect among initially possible conditioning properties of the second degree of complexity, which is $(k(k$-$1)) : 2!$.

When m is 3, we get the immediate eliminative effect among initially possible conditioning properties of the third degree of complexity, which is $(k(k$-$1)(k$-$2)) : 3!$.

Finally, when m is k we get the immediate eliminative effect among initially possible conditioning properties of the maximal degree of complexity, which is 1. Thus each positive instance of H excludes exactly one of the 2^k initially possible conditioning properties of maximal complexity from being a Necessary Condition of H.

It should be noted that the ratio of the number of immediately excluded properties to the number of initially possible conditioning properties is $1 : 2$ or $1 : 2^1$ for the 1-complex properties, $1 : 4$ or $1 : 2^2$ for the 2-complex properties, $1 : 8$ or $1 : 2^3$ for the 3-complex properties, and finally $1 : 2^k$ for the k-complex or maximally complex properties.

The additional eliminative effect of any additional positive instance of H is, as in the Simple Case, equal to the number of initially possible conditioning properties, in respect of which the additional positive instance of H differs from every one of a number of previous positive instances of H. The minimum of difference is when the additional instance differs from *some* previous instance in respect of no property from ϕ_0. The maximum of difference is, when the additional instance differs from *all* previous instances in respect of *all* properties from ϕ_0. In the minimum case the additional eliminative effect is 0, and in the maximum case it equals the immediate eliminative effect. The maximum effect, however, can occur only if there is either only one previous instance, or if all the previous instances differ from one another in respect of no one of the properties from ϕ_0.

It should be noted that even though there be no property from ϕ_0, in respect of which a certain additional instance differs from all the previous instances, the additional eliminative effect—consisting then entirely of properties of the second or higher degree of complexity—may yet be considerable. For absence of difference in respect of properties from ϕ_0 does not exclude considerable difference in respect of properties which are sums of properties from ϕ_0 and their negation-properties.

The following theorems will now be proved concerning Total Elimination and Perfect Analogy:

T1. Total Elimination on the n-level entails Total Elimination on any lower level.

Proof: For a certain n-complex initially possible conditioning property to be eliminated, means that there is some thing in which the conditioned property is present and the n-complex property absent. That the n-complex property is absent means that every one of the n properties of which it is a sum is absent. That every one of the n properties is absent means that any property which is a sum of some m of the n properties is also absent. Since any property of lower degree of complexity than n is a sum of some m of some n properties, it follows that, if *all n-complex* initially possible conditioning properties have been eliminated, then all initially possible conditioning properties of lower degree of complexity have also been eliminated.

It is clear that Total Elimination on the m-level is compatible with any state of elimination on higher levels.

T2. Perfect Analogy on the n-level entails Total Elimination on any lower level.

Proof: Consider an n-1-complex initially possible conditioning property. It is a sum of some m ($0 \leqslant m \leqslant n$-$1$) of some n-1 properties from ϕ_0 and the negation-properties of the remaining n-1-m properties. Consider a g:th property from ϕ_0 and its negation-property. The sums of the n-1-complex property, and this g:th property, and its negation-property, respectively, are two n-complex initially possible conditioning properties. The elimination of either of the n-complex properties entails the elimination of the n-1-complex property. Perfect Analogy on the n-level entails that at least one of the two n-complex properties has been eliminated. Thus Perfect Analogy on the n-level entails Total Elimination on the n-1-level which in its turn, according to *T1*, entails Total Elimination on any lower level also.

T3. Perfect Analogy on the n-level entails that no initially possible conditioning property from a higher level has been eliminated, which includes the only non-eliminated n-complex property.

Proof: That an initially possible conditioning property from a higher level includes the only non-eliminated n-complex property means, that the first property is the sum of the same n properties as the second property, and of some additional properties. It follows that the elimination of the first property would entail the elimination of the second property. Hence, by contraposition, the non-elimination of the second property entails the non-elimination of the first property.

It is clear that Perfect Analogy on the n-level is compatible with any other state of elimination on higher levels except those states which are excluded in virtue of *T3*.

T4. Absolutely Total Elimination entails Total Variation in the realm of properties in ϕ_0, and *vice versa*.

Proof: Let the number of members of ϕ_0 be k. Absolutely Total Elimination entails Total Elimination on the highest, or k-level. Total Elimination on the highest level again, according to T_1, entails Absolutely Total Elimination.

As we know, the number of initially possible conditioning properties of maximal degree of complexity is 2^k. As we also know, the immediate eliminative effect exerted by any single positive instance of the conditioned property among initially possible conditioning properties of maximal degree of complexity is 1.

Thus for the attainment of Total Elimination on the highest level, and *a fortiori* of Absolutely Total Elimination, a minimum of 2^k positive instances of the conditioned property is needed. Any two of these 2^k instances must differ in respect of at least one of the k properties from ϕ_0. This is only possible, if the 2^k instances constitute a Total Variation in the realm of the k properties. (Cf. above p. 42.)

T_5. Absolutely Perfect Analogy entails Perfect Analogy on the highest level, and *vice versa*.

Proof: Perfect Analogy on the highest level entails, according to T_2, Total Elimination on all other levels, and thus Absolutely Perfect Analogy. Perfect Analogy on any other level but the highest entails, according to T_3, that there is no Total Elimination on higher levels, and thus no Absolutely Perfect Analogy. It follows by contraposition that Absolutely Perfect Analogy entails Total Elimination on all other levels but the highest, and thus Perfect Analogy on the highest level.

Let ϕ_0 have k members. It follows from T_4 and T_5 that the minimum number of positive instances of the conditioned property needed for the attainment of Absolutely Perfect Analogy is 2^k-1.

A multitude of questions can be raised concerning the way in which the mechanism of elimination functions for the purpose of attaining various states of analogy. From the point of view of scientific observation and experimentation, two questions seem to be of particular interest:

a. What is the minimum number of positive instances of the conditioned property which are needed for the attainment of

Perfect Analogy on the n-level in combination with a maximum of elimination on (lower and) higher levels?

Let us describe the only n-complex initially possible conditioning property which is to remain uneliminated as 'the critical property.'

Perfect Analogy on the n-level entails Total Elimination on all lower levels, and is compatible with any other state of elimination on higher levels except those states which are excluded in virtue of T_3. Hence the meaning of 'maximum' is clear.

The question can be answered only on the supposition that ϕ_0 has k members.

We divide the properties in ϕ_0 into two groups. The first group consists of those n properties, of some m of which, and of the negation-properties of the remaining n-m, the critical property is a sum. The second group consists of the remaining k-n properties.

There are, in all, 2^n-1 possible ways in which the properties of the first group can be present or absent in a positive instance of the conditioned property without effecting the elimination of the critical property. (Cf. above p. 107.)

There are, in all, 2^{k-n} different ways in which the properties of the second group can be present or absent in a thing.

Thus there are, in all, $(2^n$-$1)\cdot 2^{k-n}$ or 2^k-2^{k-n} possible ways in which the properties of either group, $i.e.$, the properties in ϕ_0 can be present or absent in a positive instance of the conditioned property without effecting the elimination of the critical property.

It will not be difficult for the reader to convince himself that 2^k-2^{k-n} is the minimum number of positive instances of the conditioned property which are needed for the attainment of Perfect Analogy on the n-level, and a maximum of elimination on other levels.

If n is 1, the formula gives the minimum number of 'steps' ('observations,' 'experiments') by which we may succeed in eliminating all but one of the initially possible conditioning properties of the Simple Case, and all of the initially possible conditioning properties of the Complex Case save those which include the sole non-eliminated property of the Simple Case.

This number is 2^k-2^{k-1} or $2^k : 2$. The minimum number of things needed is thus exactly half of the minimum number needed to effect Absolutely Total Elimination. (Cf. above p. 107.)

If n is 2, the formula gives the minimum number of 'steps' by which we may succeed in eliminating all initially possible conditioning properties of the Simple Case, all 2-complex initially possible conditioning properties but one, and all other initially possible conditioning properties save those which include the sole non-eliminated 2-complex property. This number is 2^k-2^{k-2} or $(3 \cdot 2^k) : 4$ or $2^k(1 : 2+1 : 4)$. The minimum number of things needed is thus three-quarters of the number needed to effect Absolutely Total Elimination.

If n is 3, the minimum number of 'steps' is 2^k-2^{k-3} or $(7 \cdot 2^k) : 8$ or $2^k(1 : 2+1 : 4+1 : 8)$.

In general, the ratio of the minimum number of positive instances needed for attaining the Perfect Analogy in question to the minimum number needed for effecting Absolutely Total Elimination is $1 : 2^1+1 : 2^2+ \ldots +1 : 2^n$. For n equal to k, we get $1 : 2^1+1 : 2^2+\ldots+1 : 2^k$ or $(2^k$-$1) : 2^k$ which, multiplied by 2^k, gives 2^k-1, or the minimum number needed for attaining Absolutely Perfect Analogy.

β. What is the minimum number of positive instances of the conditioned property needed for the attainment of Total Elimination on the n-level?

The question can be answered only on the supposition that ϕ_0 has k members.

The criterion of Total Elimination on the n-level is that for any one of the n-complex initially possible conditioning properties, there is at least one positive instance of the conditioned property which is a negative instance of the n-complex property. That a thing is a negative instance of a sum of n properties means that it is a negative instance of every one of the n properties themselves. In general, a thing is a negative instance of more than one n-complex initially possible conditioning property. Thus the minimum number of things needed will be less than the number of n-complex initially possible conditioning properties, which is $2^n \binom{k}{n}$. What will it be?

(The qualification 'in general' means 'unless n happens to equal k.')

The 2^k maximally complex initially possible conditioning properties are divided into groups, and the groups are symmetrically arranged according to the following principle.—The first group consists of the sole property which is the sum of the k properties in ϕ_0. The last group consists of the sole property which is the sum of the negations of the k properties in ϕ_0. The number of members in each of these groups is thus 1, or $\binom{k}{0}$. The second group consists of all properties which are the sum of some k-1 properties in ϕ_0, and the negation of the remaining one property. The last group but one consists of all properties which are the sum of one property in ϕ_0, and the negations of the remaining k-1 properties. The number of members in each of these groups is k, or $\binom{k}{1}$. The third group consists of all properties which are the sum of some k-2 properties in ϕ_0, and the negations of the remaining 2 properties. The last group but two consists of all properties which are the sum of some 2 properties in ϕ_0, and the negations of the remaining k-2 properties. The number of members in each of these groups is $\binom{k}{2}$. The number of members in the group-pairs steadily increases. If k is an odd number, it reaches its maximum in the two central groups of $\binom{k}{(k+1):2}$ members each, the first of which consists of all properties which are the sum of some $(k+1):2$ properties in ϕ_0, and the negations of the remaining $(k-1):2$ properties; and the second of which consists of all properties which are the sum of some $(k-1):2$ properties in ϕ_0, and the negations of the remaining $(k+1):2$ properties. If k is an even number, it reaches its maximum in the single (asymmetrical) central group of $\binom{k}{k:2}$ members which consists of all properties which are the sum of exactly half of the properties in ϕ_0, and the negations of the remaining half.

We shall speak of these groups as the first, the second, etc., k-group.

Similarly, the $2^n\binom{k}{n}$ n-complex initially possible conditioning properties are symmetrically arranged into groups. The first group consists of all sums of some n properties in ϕ_0, and the last group of all sums of negations of some n properties in ϕ_0. The second group consists of all sums of some n-1 properties in ϕ_0 and the negation of the remaining one property, and the penultimate group consists of all sums of one property in ϕ_0 and the negation of some n-1 other properties. The number of members in the group-pairs steadily increases. If n is odd, there are two (symmetrical) central groups, and if n is even, there is only one (asymmetrical) central group.

We shall speak of these groups as the first, the second, *etc.*, n-group.

Two cases must be distinguished.

Case i. n is an even number.

A negative instance of the sole member of the first k-group is a negative instance of all members of the first n-group, and a negative instance of the sole member of the last k-group is a negative instance of all members of the last n-group. Negative instances of all members of the second k-group, but not negative instances of some of the members only, are negative instances of all members of the second n-group. Negative instances of all members of the penultimate k-group, but not negative instances of some of the members only, are negative instances of all members of the penultimate n-group. Finally, negative instances of all members of the n : 2th k-group, but not negative instances of some of the members only, are negative instances of all members of the n : 2th n-group. Negative instances of all members of the n : 2th k-group from the end, but not negative instances of some of the members only, are negative instances of all members of the n : 2th n-group from the end.

If n is not greater than k-1, then negative instances of all members of the second k-group are also negative instances of all members of the first n-group. Similarly, negative instances of all members of the penultimate k-group are also negative instances of all members of the last n-group.

Further, if n is not greater than k-2, then negative instances of all members of the third k-group are also negative instances of all members of the first and second n-groups. Similarly, negative instances of all members of the third k-group from the end are also negative instances of all members of the penultimate and ultimate n-groups.

Finally, if n is not greater than $k-n$: 2, then negative instances of all members of the n : $2+1$th k-group are also negative instances of all members of the first and second and . . . and n : 2th n-groups. Similarly, negative instances of the respective members of the $n : 2$th k-group from the end are also negative instances of the respective members of the first and second and . . . and $n : 2-1$th n-groups from the end. (It should be observed that the n-groups, counting from each end, coincide in the $n : 2$th group.)

That n is not greater than $k-n : 2$ can also be expressed as $n \leqslant 2k : 3$.

Thus, if n is not greater than $2 : 3$ of k, negative instances of the members of the $n : 2+1$th and the $n : 2$th k-groups from the beginning and the end respectively are also negative instances of the members of all n-groups.

As we know, the number of members of the two k-groups is $\binom{k}{n:2}$ and $\binom{k}{n:2-1}$ respectively.

Thus $\binom{k}{n:2}+\binom{k}{n:2-1}$ positive instances of the conditioned property, which are negative instances of the respective members of the two k-groups, will effect Total Elimination of n-complex initially possible conditioning properties. Since the number of members of the k-groups increases towards the middle of the group-arrangement, it follows that $\binom{k}{n:2}+\binom{k}{n:2-1}$ is the minimum number.

The above result is valid only on the supposition that n is 'small' relative to k, $i.e.$ not greater than $2 : 3$ of k.

Suppose that n exceeds $k-n : 2$ by 1. In this case negative instances of the respective members of the $n : 2+1$th k-group are also negative instances of the respective members of the second and . . . and $n : 2$th n-group, but not of the members of the first n-group. The smallest k-group which is such that negative instances of its members are also negative instances of the members of the first n-group, is the first k-group. It has 1, or $\binom{k}{0}$, member. Thus the minimum number of instances needed is $\binom{k}{n:2}+\binom{k}{n:2-1}+\binom{k}{0}$.

Suppose that n exceeds $k-n : 2$ by 2. In this case negative instances of the respective members of the $n : 2+1$th k-group are also negative instances of the respective members of the third and . . . and $n : 2$th n-group, but not of the members of the two first n-groups. Similarly, negative instances of the respective members of the $n : 2$th k-group from the end are also negative instances of the respective members of the second and . . . and $n : 2-1$th n-group from the end, but not of the members of the last n-group. The smallest k-groups which are such that negative instances of their members are negative instances of the two first and the last n-groups respectively, are the two first and one last k-groups respectively. Thus the minimum number of instances needed is $\binom{k}{n:2}+\binom{k}{n:2-1}+\binom{k}{1}+2\binom{k}{0}$.

Suppose that n exceeds $k-n:2$ by 3. Analogous considerations to those above would give the minimum number as $\binom{k}{n:2}+\binom{k}{n:2-1}+\binom{k}{2}+2(\binom{k}{0}+\binom{k}{1})$.

The rule for calculating the minimum number is already discernible. If n exceeds $k-n:2$ by m, then the number is $\binom{k}{n:2}+\binom{k}{n:2-1}+\binom{k}{m-1}+2(\binom{k}{0}+\binom{k}{1}+\ldots+\binom{k}{m-2})$.

The maximum value by which n can exceed $k-n:2$, is $n:2$, for which value n equals k. This limiting case corresponds to Total Elimination on the k-level, *i.e.*, to Absolutely Total Elimination. The minimum number of things needed for its attainment is, according to our formula above, $\binom{k}{k:2}+2(\binom{k}{0}+\binom{k}{1}+\ldots+\binom{k}{k:2-1})$. But, n being even, this number equals 2^k.

Case ii. n is an odd number.

The reasoning in the case where n is even can easily be modified so as to apply to this case, the treatment of which is somewhat simpler.

We have again to distinguish between the alternative that n is 'small' relative to k, meaning here that n is not greater than $(2k+1):3$, and that n is 'great' relative to k, meaning that n exceeds $(2k+1):3$.

If n is small the minimum number of instances needed for Total Elimination on the n-level is $2\binom{k}{(n-1):2}$.

If n is great and if it exceeds $k-(n-1):2$ by m, then the minimum number of instances needed is $2(\binom{k}{0}+\binom{k}{1}+\ldots+\binom{k}{m-1}+\binom{k}{(n-1):2})$.

The maximum value by which n can exceed $k-(n-1):2$, is $(n-1):2$, for which value n equals k. Thus Absolutely Total Elimination can be effected in a minimum number of $2(\binom{k}{0}+\binom{k}{1}+\ldots+\binom{k}{(k-1):2})$. But, n being odd, this number equals 2^k.

The following remarks should be added for special values of n:

If n is 1 and k is equal to or greater than 1, then Total Elimination can be effected in 2 'steps'. This is already known to us from the Simple Case.

If n is 2 and k is equal to, or greater than, 2, then Total Elimination can be effected in $k+1$ 'steps'. If, *e.g.*, k is 3, then Total Elimination of all 2-complex initially possible conditioning properties can be effected in 4 'steps'.

If n is 3 and k is equal to, or greater than, 4, then Total Elimination can be effected in $2k$ 'steps'. If, *e.g.*, k is 5, then Total Elimination of all 3-complex initially possible conditioning properties can be effected in 10 'steps'.

The cases where n is 'small' as compared with k might, from the point of view of actual observation and experimentation, be regarded as the normal cases. The more complicated formulae for cases where n is 'great' as compared with k, will certainly be very seldom relevant to actual considerations about the eliminative value of examined instances.

B. The Inverse Method.

The combinatorial calculations are the same as in the Direct Method.

The conditioned property is called H.

The (actual) conditioning properties are Sufficient Conditions of H.

Let ϕ_0 mean the same as in the Simple Case.

The initially possible conditioning properties are all properties which are the products of some m ($0 \leqslant m \leqslant n$) of some n ($1 \leqslant n$) properties from ϕ_0, and the negation-properties of the remaining $n-m$ properties.

If ϕ_0 has k members, the total number of initially possible conditioning properties is $\sum_{n=1}^{k} 2^n \binom{k}{n}$.

The Principle of Elimination is the same as in the Simple Case.

Consequently the data of elimination are afforded by negative instances of the conditioned property, which are positive instances of some of the initially possible conditioning properties.

The remaining possible conditioning properties, relative to n negative instances of the conditioned property, are defined as in the Simple Case.

If ϕ_0 has k members, the immediate eliminative effect exerted by any single negative instance of the conditioned property, amounts to $\sum_{m=1}^{k} \binom{k}{m}$ or $2^k\text{-}1$ initially possible conditioning properties.

The additional eliminative effect exerted by any additional negative instance of the conditioned property, is, as in the Simple Case, equal to the number of initially possible conditioning properties in respect of which the additional negative instance differs from every one of a number of previous negative instances. It should be noted, that even though there be no property from ϕ_0 in respect of which a certain additional instance differs from all the previous instances, the additional eliminative effect—consisting then entirely of properties of the second or higher degree of complexity—may yet be considerable.

For Total Elimination and Perfect Analogy the above theorems $T1$–$T5$ are valid. (Only in $T3$ for 'includes' we should read 'is included in.')

Suppose ϕ_0 has k members.

Then Absolutely Total Elimination and Absolutely Perfect Analogy are attainable in a minimum number of 2^k and $2^k\text{-}1$ 'steps' respectively.

The minimum number of 'steps' in which we can attain Perfect Analogy on the n-level, in combination with a maximum of elimination on (lower and) higher levels, is $2^k\text{-}2^{k\text{-}n}$.

The minimum number of 'steps' in which we can attain Total Elimination on the n-level is, again, somewhat differently calculated, according to whether n is even or odd and whether it is 'small' or 'great' relative to k.

Case ia. n is even and not greater than 2: 3 of k. The minimum number is $\binom{k}{n:2}+\binom{k}{n:2\text{-}1}$.

Case ib. n is even and greater than 2: 3 of k. The minimum number is $\binom{k}{n:2}+\binom{k}{n:2\text{-}1}+\binom{k}{m\text{-}1}+2(\binom{k}{0}+\binom{k}{1}+ \ldots +\binom{k}{m\text{-}2}))$, where m is the amount by which n exceeds $k\text{-}n$: 2. When m reaches its maximum value n: 2, i.e., when n equals k, the expression assumes the value 2^k.

Case ii$a$. n is odd and not greater than $(2k+1)$: 3. The minimum number is $2\binom{k}{(n\text{-}1):2}$.

Case ii$b$. n is odd and greater than $(2k+1)$: 3. The minimum number is $2(\binom{k}{0}+\binom{k}{1}+ \ldots +\binom{k}{m-1}+\binom{k}{(n-1):2})$. When m reaches its maximum value $(n\text{-}1):2$, *i.e.* when n equals k, the expression assumes the value 2^k.

7. The Method of Difference. The Complex Case

The conditioned property is called H.

The (actual) conditioning properties are Sufficient Conditions of H.

Let ϕ_0 mean the same as in the Method of Agreement.

The initially possible conditioning properties are, those of the initially possible conditioning properties in the Complex Case of the Inverse Method of Agreement which are present in a given positive instance of H.

Suppose that ϕ_0 has k members. Any one of the properties in ϕ_0 is either present or absent in the given positive instance of H. The number of properties which are the products of some n $(1 \leqslant n \leqslant k)$ of k properties is $\overset{k}{\underset{n=1}{\Sigma}} \binom{k}{n}$ or $2^k\text{-}1$. This therefore is the total number of initially possible conditioning properties.

The number of n-complex initially possible conditioning properties is $\binom{k}{n}$. If n is 1, this number is k (the Simple Case). If n is 2, the number is $(k(k\text{-}1)):2$. If n is k, the number is 1.

The Principle of Elimination is the same as in the Inverse Method of Agreement.

Consequently, the remaining possible conditioning properties (relative to the given positive instance and n negative instances of H) are defined as in the Inverse Method of Agreement.

The fact that no two of the initially possible conditioning properties contradict each other, makes the treatment of the Complex Case of the Method of Difference much easier than the Complex Cases of the Method of Agreement.

The immediate eliminative effect exerted by any single negative instance of H, does not have a constant magnitude. If the negative instance differs from the positive instance in respect of all properties in ϕ_0, it is wholly ineffective for the purpose of elimination. If the negative instance differs from the positive instance in respect of no property in ϕ_0, it effects Absolutely Total Elimination. 'Normally,' the immediate eliminative effect is between these two extremes.

The additional eliminative effect exerted by any additional negative instance of H, amounts to the number of initially possible conditioning properties in respect of which the additional negative instance agrees with the given positive instance and differs from *all* the previous negative instances. It should be noted that the additional negative instance may not differ from any of the previous negative instances in respect of any property in ϕ_0, and yet be of value for the purpose of elimination.

For Total Elimination and Perfect Analogy the theorems T_1-T_3 of §6 are valid. (Only in T_3 for 'includes' we should read 'is included in'.) So also is the theorem T_5 for Absolutely Perfect Analogy. Instead of T_4 we have the following theorem for Absolutely Total Elimination:

T_4'. A negative instance of the conditioned property which differs from the given positive instance of the conditioned property in respect of no property from ϕ_0, effects Absolutely Total Elimination.

(This has already been observed above.)

The following two questions will be discussed here:

a. What is the minimum number of negative instances of the conditioned property needed for the attainment of Perfect Analogy on the n-level, in combination with a maximum of elimination on (lower) and higher levels?

'Maximum' means the same as in problem a in the Complex Cases of the Method of Agreement.

The problem can be solved only on the supposition that ϕ_0 has k members.

Let there be n negative instances of H which satisfy the following two conditions:

  i. Any one of the negative instances differs from the given positive instance of H in respect of one and one only of the properties in ϕ_0.

  ii. No two of the negative instances differ from the given positive instance of H in respect of the same property in ϕ_0.

Under these conditions any less-than-n-complex initially possible conditioning property is present in some negative instance of H, and consequently eliminated. The same is true of all n-complex initially possible conditioning properties, with the one exception of the product of the n properties in which the negative instances differ from the given positive instance of H. The same is true of all more-than-n-complex initially possible conditioning properties, with the exception of those included in the only non-eliminated n-complex property. Consequently, the n things satisfying the conditions i and ii above establish a Perfect Analogy on the n-level, in combination with a maximum of elimination. It is easy to see that n is the minimum number.

If n equals k, we have the case of Absolutely Perfect Analogy, which is thus attainable in a minimum number of 'steps' equal to the number of properties in ϕ_0.

β. What is the minimum number of negative instances of the conditioned property which are needed for the attainment of Total Elimination and Perfect Analogy respectively on the n-level, in combination with a minimum of elimination on (lower and) higher levels?

'Minimum' here means all more-than-n-complex initially possible conditioning properties. Or else it means no elimination on the higher levels.

The problem can be solved only on the supposition that ϕ_0 has k members.

Let there be $\binom{k}{n}$ negative instances of H which satisfy the following two conditions:

i. Any one of the negative instances differs from the given positive instance of H in respect of all but n of the properties in ϕ_0.

ii. No two of the negative instances differ from the given positive instance of H in respect of the same properties in ϕ_0.

From i, it follows that no more-than-n-complex initially possible conditioning property can be eliminated, i.e., be present in some of the negative instances of H. From ii, it follows that any n-complex initially possible conditioning property *is* eliminated, *i.e.*, is present in some negative instance of H. The $\binom{k}{n}$ instances thus effect Total Elimination on the n-level without effecting any elimination on higher levels. It is clear that $\binom{k}{n}$ is the minimum number of instances which can achieve this.

If n is 1, the minimum number of instances needed is k. Total Elimination on the lowest level without elimination on higher levels can thus be attained in a minimum number of 'steps' equal to the number of properties in ϕ_0.

If n is k, the minimum number of instances needed is 1. This is the case of Absolutely Total Elimination which, as we already know, can be effected by a single negative instance of the conditioned property.

If we remove one of the above $\binom{k}{n}$ negative instances of H satisfying conditions i and ii, we would then have a Perfect Analogy on the n-level without having eliminated any more-than-n-complex initially possible conditioning property. It is clear that $\binom{k}{n}$-1 is the minimum number of instances which can achieve this.

If n is k, the minimum number of instances needed is $\binom{k}{k}$-1 which equals 0. Absolutely Perfect Analogy is thus attainable in 0 'steps'. This is simply another way of saying that the number of k-complex, or maximally complex, initially possible conditioning properties, is 1.

8. The Joint Method. The Complex Case

The conditioned property is called H.

The (actual) conditioning properties are Necessary-and-Sufficient Conditions of H.

Let ϕ_0 mean the same as in the Simple Case.

The initially possible conditioning properties are all the properties—the tautology and the contradiction, however, being excluded—which are presence-functions of some n ($1 \leqslant n$) properties in ϕ_0.

Suppose that ϕ_0 has k members.

The total number of initially possible conditioning properties is then $2^{(2^k)}$-2.

As we know (p. 40), a property which is a presence-function

of n properties is also a presence-function of any greater number of properties which include among themselves the n properties. Thus the initially possible conditioning properties which are presence-functions of n properties in ϕ_0, include among themselves all initially possible conditioning properties which are presence-functions of some m of the n properties. In view of this the number of initially possible conditioning properties which are presence-functions of n given properties in ϕ_0, but not of any lesser number of properties, is $2^{(2^n)}$-$2^{(2^{n-1})}$.

n properties can be selected from among k properties in $\binom{k}{n}$ different ways. Thus the number of initially possible conditioning properties which are presence-functions of some n properties in ϕ_0, but not of any lesser number of properties, *i.e.*, the number of n-complex initially possible conditioning properties, is $\binom{k}{n}(2^{(2^n)}$-$2^{(2^{n-1})})$.

If n is 1, we get the initially possible conditioning properties of the Simple Case. Their number, as we already know, is $2k$.

If n is 2, we get the 2-complex initially possible conditioning properties. Their number is $6k(k$-$1)$. They are thus three times as many as in the Complex Cases of the Method of Agreement. (Cf. above p. 103.)

If n is k, we get the maximally complex initially possible conditioning properties. Their number is $2^{(2^k)}$-$2^{(2^{k-1})}$.

It should be noted that $\sum\limits_{n=1}^{k} \binom{k}{n}(2^{(2^n)}$-$2^{(2^{n-1})})$ equals $2^{(2^k)}$-2.

The Principle of Elimination is the same as in the Simple Case.

Consequently, the data of elimination are afforded by positive instances of the conditioned property which are negative instances of some of the initially possible conditioning properties, and by negative instances of the conditioned property which are positive instances of some of the initially possible conditioning properties.

The remaining possible conditioning properties, relative to m positive and n negative instances of the conditioned property, are defined as in the Simple Case.

As in the Simple Case, the idea underlying the use of the Joint Method in the Complex Case is to compare a set of remaining possible Necessary Conditions, and a set of remaining possible Sufficient Conditions, of the conditioned property. The common members of both sets will constitute a set of remaining possible Necessary-and-Sufficient Conditions.

As in the Simple Case, there are two forms of the Joint Method in the Complex Case also. We call them, as before, the Double Method of Agreement and the Joint Method of Agreement and Difference.

We shall not deal separately with the two sub-methods for the Complex Case. As was shown, their employment in the Simple Case necessarily leads to concordant results when applied to the same 'material of observation or experimentation.' (Cf. above p. 102.) The proof applies immediately to the Complex Case also.

As in the Simple Case, the effect of elimination depends upon two factors, *viz.*, variation among the members of a set of positive instances and a set of negative instances of the conditioned property, taken separately; and resemblance between the members of two such sets taken jointly. The resemblance-effect is independent of the variation-effect and responsible for the total eliminative effect. (Cf. above p. 100).

For Total Elimination and Perfect Analogy none of the theorems T_1-T_5 of §6 are valid. Instead we have the following theorems:

T_1'. Total Elimination on the n-level is compatible with, but does not entail, Total Elimination on any lower level.

Proof: Let there be a property which is a presence-function of n properties but not of any m ($1 \leqslant m < n$) of them. Let there be another property which is a presence-function of some m ($1 \leqslant m < n$) of the n properties. (This second property is then also a presence-function of the n properties.) Consider the combinations of presence-values in the n properties, for which the first property is present in a thing, and those for which it is absent. Suppose that the second property were present for exactly the same combinations and absent for exactly the same combinations. This would entail that the two properties were identical, and hence that the first property was a presence-function of the same m properties as the second property, which is contrary to our assumptions. Thus there must exist at least one combination of presence-values in the n properties, for which either the first property is present and the second absent, or the first absent and the second present. In other words: either the presence of the first property in a thing must be compatible with the absence of the second property in it, or the absence of the first property must be compatible with the presence of the second.

That a certain n-complex initially possible conditioning property is eliminated, means that either there is a thing in which the conditioned property is present and the n-complex property absent, or one in which the conditioned property is absent and the n-complex property present. The elimination of an n-complex property never entails the elimination of a less-than-n-complex property, since either the presence of the n-complex property is compatible with the absence of the less-than-n-complex property, or its absence is compatible with the presence of the less-than-n-complex one. For the same reason, however, the elimination of any given n-complex property is compatible with the elimination of any given less-than-n-complex property. Thus, in particular, the elimination of all n-complex properties is compatible with, but does not entail, the elimination of all less-than-n-complex properties also.

$T_2{}'$. Perfect Analogy on the n-level is compatible with, but does not entail, Total Elimination on any lower level.

Proof: That a certain n-complex initially possible conditioning property is not eliminated, means that there is no thing in which it is either present in the absence of, or absent in the presence of, the conditioned property. Since either the presence of the n-complex property is compatible with the absence of any given less-than-n-complex property, or its absence with the presence of the less-than-n-complex one, it follows that the non-elimination of the n-complex property is compatible with, but does not entail, the elimination of the less-than-n-complex property.

$T_3{}'$. Perfect Analogy on the n-level is compatible with, but does not, entail, Total Elimination on any higher level.

This follows immediately from the considerations which led us to $T_1{}'$ and $T_2{}'$.

T_4'. A negative and a positive instance of the conditioned property which do not differ in respect of any of the properties in ϕ_0, effect Absolutely Total Elimination.

Proof: Consider an initially possible conditioning property. In order not to be eliminated, this property must be present in the positive instance and absent in the negative instance, of the conditioned property. The initially possible conditioning property is a presence-function of some n properties in ϕ_0. Each one of these n properties has the same presence-value in the positive, and in the negative, instance of the conditioned property. Since the initially possible conditioning property is a presence-function of the n properties, it is either present or absent for the combination of presence-values which the n properties happen to have in the two instances. If it is present, it must be present in both instances, and if it is absent it must be absent in both instances, of the conditioned property. It cannot be present in the positive and absent in the negative instance. Hence it must be eliminated.

(This theorem is the same as T_4' of §7.)

T_5'. Absolutely Perfect Analogy is attainable on any level of complexity.

This follows immediately from the fact that Perfect Analogy on the n-level is compatible with Total Elimination on any lower level (T_2') and any higher level (T_3').

That Absolutely Perfect Analogy is attainable on any level, constitutes a great advantage of the Joint Method over the Method of Agreement and the Method of Difference. The Joint Method is the only method of elimination where it is possible to 'isolate' one single property of any degree of complexity, (in a certain set ϕ_0), and to maintain that no other property of lower or higher degree of complexity can actually be a conditioning property of the conditioned property in question. In the Method of Agreement and the Method of Difference the corresponding result is possible only for the relatively unimportant and 'trivial' case, where the sole non-eliminated property is itself of maximal complexity.

The following question remains to be discussed:

What is the minimum number of instances needed for the attainment of Absolutely Perfect Analogy on the n-level?

The problem can be solved only on the supposition that ϕ_0 has k members.

We shall call the only non-eliminated n-complex property 'the critical property.'

Consider the perfect disjunctive and conjunctive normal denotations of the critical property in terms of the names of the k properties of ϕ_0. Tautology and contradiction being excluded, the first is a 1- or . . . or 2^k-1-termed disjunction-name of k-termed conjunction-names. The second is a 1- or . . . or 2^k-1-termed conjunction-name of k-termed disjunction-names. Let the first normal denotation be an m-termed disjunction. In consequence of the complementary nature of the normal forms (p. 42), the second will then be a 2^k-m-termed conjunction.

Let there be m positive instances of H satisfying the following two conditions:

i.   Each one of the positive instances is also a positive instance of some of the m k–termed conjunctions, of which the critical property is a disjunction.

ii.   No two of the positive instances are positive instances of the same k-termed conjunction.

Further, let there be 2^k-m negative instances of H satisfying the following two conditions:

i'.   Each one of the negative instances is also a negative instance of some of the 2^k-m k-termed disjunctions, of which the critical property is a conjunction.

ii'. No two of the negative instances are negative instances of the same k-termed disjunction.

Consider next one of the initially possible conditioning properties other than the critical property. Its perfect disjunctive normal denotation in terms of the names of the k properties in ϕ_0, is either a less-than-m-termed, or an m-termed, or a more-than-m-termed disjunction-name of k-termed conjunction-names. Accordingly, its perfect conjunctive normal denotation is either a more-than-2^k-m-termed, or a 2^k-m-termed, or a less-than-2^k-m-termed conjunction-name of k-termed disjunction-names.

Suppose the normal disjunctive denotation to be less-than-

m-termed or m-termed. Then at least one of the m-positive instances of H which satisfy the conditions i and ii above, is a negative instance of the initially possible conditioning property in question. For otherwise, all the m positive instances of H would also be positive instances of at least one of the less-than-m, or m k-termed conjunctions, and this would contradict conditions i and ii. Hence any initially possible conditioning property, (other than the critical property), the perfect disjunctive normal denotation of which is a less-than-m-termed or m-termed disjunction of k-termed conjunctions, is eliminated.

Suppose next the normal disjunctive denotation to be (m-termed or) more-than-m-termed. Then at least one of the 2^k-m negative instances of H which satisfy the conditions i' and ii' above, is a positive instance of the initially possible conditioning property in question. For otherwise, all the 2^k-m negative instances of H would also be negative instances of at least one of the (m or) less-than-2^k-m k-termed disjunctions of which the property in question is a conjunction, and this would contradict conditions i' and ii'. Hence any initially possible conditioning property, (other than the critical property), the perfect disjunctive normal denotation of which is an (m-termed or) more-than-m-termed disjunction of k-termed conjunctions is eliminated.

Consequently, the m positive, and the 2^k-m negative instances of the conditioned property establish an Absolutely Perfect Analogy on the n-level. The sum of m and 2^k-m is 2^k. It is not difficult to see that 2^k must be the minimum number.

It is interesting to observe that the minimum number depends only on the number k of properties in ϕ_0, and not on the degree n of complexity in the critical property. The minimum number is the same for all degrees of complexity. What varies is the number of positive and negative instances, respectively, of the conditioned property. Not even these numbers, however, depend upon the degree of complexity n, as here defined, but upon the number (m) of terms in the perfect disjunctive normal denotation, (in terms of the names of the k properties in ϕ_0), of the critical property. (This number m could also be used to measure the degree of complexity of the initially possible conditioning properties. Such a measure would, however, be somewhat awkward, considering that the k properties in ϕ_0, and their negation-properties, would then themselves be 2^{k-1}-complex.)

9. Elimination and the Practice of Science

The four sub-methods of the general Method of Elimination exhaust the ways in which the eliminative mechanism may operate.

The four standard questions corresponding to the four sub-methods described, do not, however, exhaust the multitude of problems raised in actual scientific investigation into nomic connexions. In order to understand the nature of the eliminative procedure in science in its full concreteness, we should always bear in mind that the types of problem here treated with the instruments of logic represent isolated and simplified aspects of situations which will, as a rule, be of a highly complicated structure. On the other hand, those aspects provide, so to speak, the logical 'atoms' out of which the more complicated situations can be reconstructed.

We shall not work out here the reconstruction of actual cases from the practice of science. But let us briefly consider some patterns for a combined use of the various inductive methods. We shall distinguish the following types of complication:

i. Problems of conditioning relations which are more general than the types of question treated under the head of the four methods.

The most general question as to nomic connexion which deserves attention would seem to be this: What nomic connexions prevail among the members of a set of properties ϕ_0? The question is attacked by inquiring, with regard to any pair of presence-functions of members of ϕ_0, what criteria for invalidating laws can be applied in given contexts of things.

From an initial question of this general character there will frequently emerge, in the course of investigation, questions of a more determinate nature, such as: What factors (properties) are connected with a specified factor (conditioned property)? This question, being a disjunction of all (or some) of the standard questions treated under the head of the four methods, is attacked by successively applying the eliminative principles of the respective methods to the case under consideration.

The next stage would then be the emergence of the four questions themselves. This, however, is not the last stage into which such a situation may develop. We have also to consider:

ii. Problems of conditioning relations which are more specific than the types of question treated under the head of the four methods.

Thus we might, *e.g.*, select two factors (properties) H and A and ask whether they are connected by law in a specific way. We examine, say, a number of positive instances of H and find that, in spite of much variation in the accompanying circumstances, A is a common feature of them all. Suppose, however, that it is impracticable for us to produce a still greater variation in the instances, whereby to increase our confidence in the assumption that A really is a Necessary Condition of H. We may then resort to negative instances of A, and examine them as to the presence or absence of H. The 'typical' case, known from scientific experimentation, occurs when the change-over from positive instances of H to negative instances of A takes the form of a *removal* of A from the original positive instances of H. If, so far as our experience goes, this removal invariably makes the nature of the instances of H turn over from positive to negative, then, in so far as the accompanying circumstances remain unaffected by the removal, we have reason to believe that H is a Sufficient Condition of A. Since the law that H is a Sufficient Condition of A is logically identical with the law that A is a Necessary Condition of H, we have in this case supported the same law by resort to two different methods of elimination.

iii. Problems which do not directly involve a question of conditioning relations, but rather of finding Contributory Conditions, Indispensable Contributory Conditions, Counteracting Conditions or Substitutable Requirements, of a given property.

iv. Problems of conditioning in which the conditioned property is viewed as a presence-function of some other properties.

It should be observed, that if the conditioned property is a disjunction, then positive instances of it can be provided in as many different ways as the disjunction has terms. Similarly, if the conditioned property is a conjunction, negative instances of it can be provided in as many ways as there are terms in the conjunction.

Chapter Five

INDUCTION AND DEDUCTION

1. The Supplementary Premisses of Induction

INDUCTIVE reasoning, as it stands, is non-demonstrative. It is an old idea that what entitles us to speak of induction as 'inference' is the fact that in inductive reasoning certain premisses are suppressed which, in combination with the stated premisses, would make the argument conclusive or demonstrative. Or as Mill says, following Archbishop Whateley, 'Every induction may be thrown into the form of a syllogism, by supplying a major premiss.'

We shall call the stated or explicit premisses of an inductive inference (of the second order) the instantial premisses.[1] We shall call the suppressed or implicit premisses, needed for making the argument conclusive, the supplementary premisses. The latter are sometimes also called the 'presuppositions' of induction.

In Chap. III, §1 the nature of the instantial premisses was made clear.—It should be observed that in Induction by Elimination the instantial premisses of the argument are not, strictly speaking confirming instances of the conclusion but data of elimination (p. 87), *i.e.*, disconfirming instances of 'concurrent' conclusions. However, the things which afford the disconfirming instances of these 'concurrent' conclusions also afford confirming instances of the conclusion itself.

Whether the supplementary premisses really are at the back of the reasoner's mind, when he draws inductive conclusions, need not be discussed here. The idea that they are, or should be (!) present seems to me an unrealistic construction inspired

---

[1] The term 'instantial premiss' I have taken from W. E. Johnson.

by the deep-rooted prejudice that sound reasoning must always be syllogistic or deductive. From the point of view of understanding the nature of human knowledge it is desirable that this deductivistic prejudice should be given up and that the existence of 'genuine' species of reasoning other than the demonstrative should be acknowledged. (Cf. above p. 21).

But even if we do not subscribe to the view that induction, *qua* inference, is deduction from partially suppressed premisses, it is an important task of the Logic of Induction to make clear the nature of the supplementary premisses.

In Chap. IV, §1 we distinguished between two chief Methods of Induction: Induction by Enumeration and Induction by Elimination.

In the case of Induction by Enumeration one supplementary premiss is needed. It states, broadly speaking, that if some things afford confirming instances of a certain inductive conclusion, then all things afford confirming instances of this conclusion. We might refer to this premiss as the Postulate of the Uniformity of Nature. It is a very sweeping principle. Further discussion of its more precise content will not concern us here.

In the case of Induction by Elimination two supplementary premisses are needed.

The first premiss is of an existential nature. It states the existence of actual conditioning properties of a certain conditioned property. We shall call it the Deterministic Postulate.

The second premiss is concerned with the selection of the initially possible conditioning properties (the set ϕ_0). It states two things: first, that the range of initially possible conditioning properties includes the actual conditioning properties, and secondly, that the state of analogy (p. 91) among any given number of things in respect of the initially possible conditioning properties, can be settled on the basis of an enumeration of the data of elimination. We shall call it the Selection Postulate.

Attempts to prove the truth of the supplementary premisses independently of, or prior to, the inductive conclusions which they are to support, are well-known from the history of the subject.

Here we are not directly interested in the problem of truth. Our primary interest is in what may be called the problem of reconstruction, *i.e.* laying down the exact content of the supplementary premisses which, in combination with the instantial premisses, make inductive inference demonstrative. The second problem has traditionally been neglected at the expense of the first.

The reconstructive task could also be described as one of determining the 'distance' separating induction from demonstrative inference. There obviously has been, and is, a tendency to regard the distance as comparatively short, *i.e.*, to regard the supplementary premisses as relatively modest principles. Further there has been, and is, a tendency to regard the distance as fairly uniform, *i.e.*, to assume the exact content of the supplementary premisses as the same, if not for all, at least for large groups of inductive inferences.

Now the reconstruction of the supplementary premisses clearly shows that these two presuppositions as to the distance separating induction from deduction are mistaken. The premisses are far more sweeping than is apparent from a superficial glance. Further, and perhaps even more important, the premisses are not the same from case to case; their exact content essentially depends upon the individual nature of the case to which they are applied.

It is evident that attempts to prove the truth of the supplementary premisses of induction have been greatly encouraged by the two presumptions which our reconstruction shows to be mistaken. Once the reconstructive undertaking has been accomplished, it is difficult to see how these attempts could be continued with any reasonable hope of success. The reconstruction thus becomes indirectly relevant to the question of proving the premisses. The task before us will be a most instructive lesson in the limitations of the power of the human intellect to advance by argument 'beyond the evidence of our memory and senses.'

2. *The Deterministic Postulate*

This supplementary premiss chiefly occurs under the name of the Principle of Determinism or Law of Universal Causation. Its meaning, in the various authors who have employed it, is usually obscure. This is so for two chief reasons. First they have not clearly distinguished between the different kinds of logical conditioning which are inherent in the popular notions of causality and law. Secondly, they have not clearly realized that causal determination is relative to a set of properties within a Universe of Properties.

In Chap. III, §2 we suggested the following (partial) analysis of the Principle of Determinism in terms of conditional relationship:

Every member of a certain Universe of Properties is a Determined Property in the set of all members, of which it is logically totally independent.

If the Deterministic Postulate is understood in the suggested way, it would follow (p. 72 f.) that for every property of a certain universe there is a set of initially possible conditioning properties which contains at least one Necessary Condition, at least one Sufficient Condition, and at least one Necessary-and-Sufficient Condition of the property. The postulate would thus enable us to conclude from a state of Absolutely Perfect Analogy among things in respect of this range, to the truth of a particular Law of Nature.

We shall not here discuss to which Universes of Properties the postulate could possibly be applied.

It should be observed that the above form of the Deterministic Postulate is the weakest which satisfies the following two conditions:

  i. Its content is independent of whatever sub-method of the general Method of Exclusion is being used.
  ii. Its content is independent of the individual nature of the conditioned property.

It seems that any form of the Deterministic Postulate which can claim for itself an 'intuitive plausibility' must satisfy at least these two conditions.

It is easy to see that the above weakest form of the Deterministic Postulate ceases to be effective as a supplementary premiss of induction as soon as we have to reckon with Plurality or with Complexity of Conditions.

Let us first consider Plurality of Conditions.

If the given conditioned property has several conditions of a certain kind, then a state of Absolutely Perfect Analogy can never be attained. The utmost attainable by means of elimination is a reduction of the set of initially possible conditioning properties so that it contains only the actual conditioning properties of the given conditioned property. The attainable state of analogy is thus determined by the number n of actual conditioning properties.

In order to conclude from a state of analogy, other than an Absolutely Perfect Analogy, to the truth of particular Laws of Nature we must rely on a stronger form of the Deterministic Postulate. It states:

A given member of a certain Universe of Properties has (at least) n Sufficient (Necessary, Necessary-and-Sufficient) Conditions in the set of members of which it is logically totally independent.

One and the same member of the universe may have a different (minimum) number of the different kinds of condition, *i.e.*, of Sufficient, Necessary, and Necessary-and-Sufficient Conditions. If this is the case, the stronger form of the Deterministic Postulate does not satisfy condition i above, *i.e.*, its content varies according to which sub-method of the general Method of Elimination is used.

Different members of the universe may have a different (minimum) number of conditions of the same kind. If this is the case, the stronger form of the Deterministic Postulate does not satisfy condition ii above, *i.e.*, its content varies according to the choice of the conditioned property.

Thus only in cases which must be regarded as 'exceptional' rather than 'normal' does the stronger form of the Deterministic Postulate satisfy the conditions i and ii which seemed to be minimum requirements, as it were, if the postulate is to claim 'intuitive plausibility' in its favour.

Let us next consider Complexity of Conditions.

As will be remembered, a Perfect Analogy on the n-level as to the initially possible Necessary Conditions of a given conditioned property entails, that no property of superior complexity has been eliminated, which itself includes the sole non-eliminated n-complex property. Thus a Perfect Analogy on the

n-level is an Absolutely Perfect Analogy only if n happens to be the maximum degree of complexity.

It follows, that in the case of Necessary Conditions, the weakest form of the Deterministic Postulate can effectively be used as a supplementary premiss of induction only for maximally complex conditioning properties.

In order to remove this limitation a stronger form of the Deterministic Postulate is needed. *Its* weakest form would run:

A given member of a certain Universe of Properties has at least one Necessary Condition of not higher degree of complexity than n in the set of members of which it is logically totally independent.

This form of the postulate would enable us to conclude from a state of Perfect Analogy on the n-level to the truth of a particular Law of Nature.

What has been said here of Necessary Conditions applies, *mutatis mutandis*, also to Sufficient Conditions.

Necessary-and-Sufficient Conditions are, however, in a somewhat different position. As will be remembered, a Perfect Analogy on the n-level, as to initially possible Necessary-and-Sufficient Conditions of a given conditioned property, may, for any n, be an Absolutely Perfect Analogy.

It follows, that in the case of Necessary-and-Sufficient Conditions, the weakest form of the Deterministic Postulate may effectively be used as a supplementary premiss of induction.

One and the same member of the Universe of Properties may have a different minimum degree of complexity for its various kinds of condition, *i.e.*, for its Necessary, Sufficient, and Necessary-and-Sufficient Conditions. If this is the case, the required form of the Deterministic Postulate does not satisfy condition i above.

Different members of the Universe of Properties may have a different minimum degree of complexity for conditions of the same kind. Some members may have simple or moderately complex conditions, others only highly complex ones. If this is the case, the required form of the Deterministic Postulate does not satisfy condition ii above.

As regards 'intuitive plausibility' the stronger form of the Deterministic Postulate is thus in a similar position when we consider complexity of conditions as when we consider their plurality.

It should be added, that in normal scientific practice we have to reckon with plurality rather than singularity, and with complexity rather than simplicity of conditions. This means that the weakest form of the Deterministic Postulate, or that form which may be viewed as a reasonable approximation to what is commonly known as the Law of Universal Causation, is practically useless as a supplementary premiss or 'presupposition' of induction.

All forms of the Deterministic Postulate are relative to sets of properties within a Universe of Properties. Unless the number of members of these sets is known, all forms of the Deterministic Postulate are undecidable, *i.e.*, neither verifiable nor falsifiable, on the basis of particular nomic connexions.

3. *The Selection Postulate*

As said in §1, the Selection Postulate is a combination of two conditions, *viz.*

  i. that the set of initially possible conditioning properties includes the set of actual conditioning properties, and,
  ii. that the state of analogy among any given number of things in respect of the initially possible conditioning properties can be settled on the basis of an enumeration of the data of elimination.

It is not difficult to select the range of initially possible conditioning properties so that it fulfils each of the two conditions separately. The difficulty is to find a range which satisfies them jointly.

The first condition is fulfilled if the initially possible conditioning properties are those properties in the universe which are logically totally independent of the given conditioned property.

The second condition is fulfilled if the number of initially possible conditioning properties is known.

It follows that both conditions can be simultaneously fulfilled if either, α. for any given conditioned property, the number of properties in the universe, of which it is logically totally independent, is known, or,

β. for any given conditioned property, a set with a known number of members can be known to include the set of actual conditioning properties.

Under alternative α) we shall talk of an inclusive, and under alternative β) of an exclusive, form of the Selection Postulate.

For the inclusive form of the postulate Keynes introduced the name 'Postulate of Limited Variety.'

What can be said in defence of an inclusive form of the postulate?

An important realm of inductive inquiry is the realm of sense-qualities and properties which are presence-functions of sense-qualities. It is sometimes thought that the existence of so-called perception-thresholds makes the number of sense-qualities finite. The eye, it is said, cannot distinguish more than a finite number of shades of colour, and the ear a finite number of pitches of sound. These facts, however, have no bearing on the Postulate of Limited Variety as a supplementary premiss of induction, if, as is at least highly plausible, we regard different shades of colour or pitches of sound as mutually exclusive, and thus not logically independent. For then any point of visual space-time will possess, for formal reasons, one and only one shade of colour, and any point of auditory space-time one and only one pitch of sound, quite irrespective of whether the number of shades or pitches, as such, is finite or not. This means that the variety with which the postulate is concerned is, so to speak, in a different 'dimension' from the variety to which the argument from thresholds sets a limit. What matters from the point of view of the postulate is, in the terminology of W. E. Johnson, not whether the number of different sense-quality determinates—such as particular shades of colour, pitches of sound, kinds of feeling, volition, *etc.*—is finite or not, but whether the different determinables in the realm of sense-perception—such as the three so-called dimensions of colour and of sound, the so-called dimensions of feeling, of volition, *etc.*—are limited or unlimited in number. I do not know of any argument, analogous to that from the existence of thresholds, which would be in favour of the finitude of these determinables.

We need not here continue the discussion of the finite or infinite character of the various realms of properties in which inductive inquiry is pursued. The reason for this is the following:

The finitude of the ranges, *i.e.*, the existence of a cardinal number for them, though necessary, is not sufficient for the effective use of an inclusive form of the Selection Postulate as a supplementary premiss of induction. If the state of analogy is to be settled on the basis of an enumeration of data of elimination, we must assume the cardinal number of the ranges not only to exist but also to be known. It is, however, difficult to imagine any serious argument which could be alleged in support of this stronger assumption of finitude for any Universe of Properties where induction is pursued.

For the exclusive form of the postulate we may use the name 'Postulate of Irrelevance.'

It is an incontestable and important fact that in most cases of inductive inquiry we look for the conditions of a given property among a comparatively small number of properties, though we are perfectly conscious of the presence of a great number of properties which we neglect. The properties, in other words, fall initially into two groups, *viz.*, possibly relevant factors and irrelevant ones. The latter are excluded without further ado from the initial set of possible conditioning properties.

One of the reasons for regarding certain properties as initially irrelevant to a scientific observation or experiment is simply the testimony of previous experience. The terms 'previous experience' should here be taken in a wide sense, comprising a large bulk of pre-scientific, more-or-less impersonal and unsystematic records of eliminative data registered during the history of the human race. To take only one example: There hardly exists any systematic refutation of astrological beliefs. But the scientific attitude is—in most cases at least—entitled to neglect them for the purpose of ascertaining causal connexions in the lives of human individuals, and to regard the neglect as founded, not on unwarranted prejudice, but on solid and long experience of the way in which factors of the world are connected as conditions of one another.

Besides experiential or material reasons for irrelevance, there seem also to exist conceptual or formal reasons. One instance of the latter is connected with the distinguishing of Logical Types or Universes of Properties and Things. This means that the irrelevance of certain factors to certain other factors is explained by showing that the factors actually belong to different universes, and therefore cannot be nomically connected.

We shall here mention one case only, in which an appeal to the distinction of Logical Types can be made:

Maxwell appears to have been the first to state clearly the important principle that the validity of Laws of Nature is never restricted by space and time *as such*. Or to put it more precisely: that spatio-temporal co-ordinates of things are never Sufficient or Necessary or Necessary-and-Sufficient Conditions for the occurrence of a certain property in the things. We shall call this principle the Postulate of Spatio-Temporal Irrelevance.

Is this postulate valid, and what are the reasons or evidence for its validity? Some authors take the view that the postulate is itself a generalization from experience and that it 'has no justification in pure logic'.[1] This attitude, however, is hardly sound. It seems much more plausible to think that the Postulate of Spatio-Temporal Irrelevance is concerned with matters of logic than with matters of fact. The question of logic involved may, moreover, be one of distinguishing types. It could be suggested that positions in space and time are not 'properties' which we attribute to physical things—though we frequently use a loose mode of expression which makes us believe that they are—but that the physical things themselves are space-time-volumes. To say of a certain thing that it is phosphorus, or that it is red, is to say that certain properties are characteristic of a certain place in space and time. It could also be suggested that positions in space and time are not properties, but properties (relations) of properties of physical things. In either case, space and time on the one side, and properties of physical things on the other, belong to different universes and this would explain why they cannot be connected by law as mutually conditioning one another.

[1] Jeffreys, *Theory of Probability* (1939), p. 11.

INDUCTION AND DEDUCTION

We need not here continue the discussion concerning the reasons for regarding properties as initially irrelevant. It will suffice to make the following observations:

The existence of reasons for regarding properties as initially irrelevant is, though necessary, not sufficient for the effective use of an exclusive form of the Selection Postulate as a supplementary premiss of induction. If the state of analogy is to be determined on the basis of an enumeration of data of elimination, we must know, not only which properties are initially irrelevant, but also that those properties are *all but k* of the members of the universe, of which the given conditioned property is logically totally independent. It is, however, difficult to imagine any serious argument which could be alleged in support of this stronger assertion of irrelevance for any Universe of Properties where induction is pursued.

*   *   *   *   *

Taken together, the two supplementary premisses which bridge the gap between induction and deduction, *i.e.*, which in combination with the data of elimination make inductive inference conclusive or demonstrative, run as follows:

The conditioned property H has, among k initially possible conditioning properties, (at least) m (not-more-than-)n-complex conditions of a certain kind. Here the values of the three coefficients k and m and n would have to be determined for each case separately, and would depend partly upon which sub-method of the general Method of Elimination is used, and partly upon the choice of the conditioned property H.

It should be observed that this joint form of the supplementary premisses of induction is a falsifiable proposition in the sense that its refutation would follow from the elimination, among the k initially possible conditioning properties, of more than k-m properties (of any degree of complexity). This should suffice to show that the attempt to prove the presuppositions or supplementary premisses of induction independently of, or prior to, the several inductive conclusions which they are to support, is an idle undertaking.

Chapter Six

INDUCTION AND DEFINITION

1. Actual and Ideal Induction

THE use of induction will be called 'ideal,' if it is possible to come to know whether any given thing does or does not afford a confirmation of a given law.

The actual use of induction is certainly not always ideal. It may indeed be questioned whether the ideal use of induction is not a limit to be approached, but never reached, in practice. It is important to recognize this for a complete account of inductive method, though it may be useful to ignore possible imperfections for the purpose of a first description of the inferential mechanism of induction.

Thus induction is not ideal, if there are difficulties over the verification and falsification of propositions to the effect that a certain thing affords a confirming instance of a certain law. Of such difficulties there are, moreover, various types. Here we mention some only:

First, there are the difficulties constituting the general problem as to whether an 'ultimate' verification or falsification of material propositions is ever possible. (Cf. above p. 37.)

Secondly, there are difficulties arising from the fact that the instances of a law may themselves be Universal or Existential Propositions.

Thirdly, there are difficulties connected with the definition of the properties connected by law in an inductive conclusion. It is clear that difficulties over definition may cause difficulties in deciding whether a thing does or does not afford a confirmation of a law.

In this book we shall discuss only those imperfections in the use of induction which arise from difficulties over definition. Of such difficulties there seem to be two principal types:

Sometimes there is uncertainty as to whether the nomically related properties are or are not logically independent. Sometimes again there is uncertainty as to the way in which the nomically related properties are presence-functions of members of some set of properties ϕ_0.

The first imperfection is removed as the formation of well-defined scientific notions advances. The removal of the second imperfection is attained through a rectification of defective formulations of the laws themselves.

2. Induction and the Formation of Concepts

Let us consider an example which has been frequently discussed in the literature of induction and scientific method:[1]

We know from chemistry that the melting-point of phosphorus is approximately 44° C. We arrive at this result in what may be termed an inductive way, *i.e.*, we melt different pieces of a substance known to us under the name of phosphorus, and find that they all melt at the same temperature of 44° C. if the experiment has been carefully performed. From these observations we conclude that *all* pieces of phosphorus melt at 44° C., or, as we may also express it, that melting at 44° C. is a Necessary Condition of a thing being phosphorus. This is a Universal Implication $H \subset A$, if H denotes the property 'phosphorus,' and A the property 'melting at 44° C.'

We shall for the moment abstract from the obvious fact that the presence of H cannot in itself be sufficient for the presence of A, but that the Law of Nature in question, when formulated in full, is of the structure: Supposing such and such conditions (of pressure, *etc.*) to be fulfilled, then phosphorus melts at 44° C. The relevance of this point will be estimated later (§3).

Suppose then that we find a thing which is similar to the previous ones, but which does not melt at 44° C. Does this fact mean that our conclusion as to the melting-point of phosphorus is falsified?

[1] Cf. *The Logical Problem of Induction*, pp. 48–52.

Obviously, what has happened could be regarded as a falsification of the law $H \subset A$. But there is also another way left open and in the practice of science under similar circumstances it is very often resorted to. We simply raise the question whether the melted thing was phosphorus or not. This leads us to consider the definition of phosphorus. The consideration may take roughly the following course:

In order to arrive at the generalization $H \subset A$ it was necessary to have criteria enabling us to decide when it was a piece of phosphorus we were dealing with, and when it was not. It is quite conceivable that we are not able to enumerate exactly the criteria we used in our experiments, and a scientist in the first place would hardly bother about such an enumeration. But certainly we have relied upon *some* criteria, since we have chosen a definite kind of substance for the experiments. Let us assume these criteria to have been H_1, H_2, and H_3, *e.g.*, macroscopic criteria such as a certain colour, smell, and taste.

In enunciating the law about the melting-point we were in the first place enunciating a discovery, *viz.*, that pieces of a substance with the properties in question have been found to exhibit a further property, the property of melting at a constant temperature. In so far as the generalization $H \subset A$ is to mean that, whenever in the future we find a substance with the characteristics H_1, H_2, and H_3, it will exhibit the further discovered property also, then this generalization has actually been refuted by experience.

Although the word 'phosphorus' was used at the outset for things exhibiting the macroscopic properties mentioned, it is by no means certain that we wished, even at the beginning, to *define* phosphorus in terms of H_1, H_2 and H_3. We need not have expressed any opinion at all as to what phosphorus 'really' is. We simply asserted that the substance was phosphorus, as though phosphorus were something fixed and given, about whose definition there was no reason to bother. A certain coexistence of a number of easily observable properties had made us familiar with something called phosphorus, and which of these properties were defining and fundamental ones and which again connected with phosphorus by law, was a question which had never before occurred to us. The first time that we were

confronted with it was in the above situation, where the pro-
properties H_1, H_2 and H_3, which we had been accustomed to
regard as reliable signs for the presence of phosphorus, were
present, but a further property which had hitherto always
accompanied them was absent.

Now the question occurs: what then *is* phosphorus? Is the
thing under examination phosphorus or not? In such a situation
it is quite conceivable that we explicitly renounce every pre-
tension of regarding the properties H_1, H_2 and H_3 as the decisive criteria for what may be called phosphorus, and accept
the experimentally discovered accompanying property as more
fit for that purpose.

This, however, does not necessarily mean that we explicitly
announce the melting-point itself as a defining property
(among others) of phosphorus. We *may* do so, but there is also
another possibility to be considered.

In making our decision as to whether a positive instance of
$H_1\&H_2\&H_3$ not melting at 44° C. is, or is not, a falsification of
the law of the melting-point, we have regard to a multitude of
circumstances. It is reasonable to assume that among those
circumstances are to be found assumptions which are themselves inductive conclusions. For example, we may assume that
if a positive instance of $H_1\&H_2\&H_3$ does not melt at 44° C.,
then it differs from phosphorus also in other properties, *e.g.*, its
microphysical structure, and that this difference is the 'cause'
of some positive instances of $H_1\&H_2\&H_3$ melting at the temperature in question and others not doing so.

Thus we can let the law $H \subset A$ escape invalidation, not
because the melting-point is itself one of the defining criteria of
phosphorus, but because it indicates the presence of another
property that explains why phosphorus melts at exactly this
temperature. Here it is important to observe that even if we do
not know of any such property, or even if every property within
our present experience can be shown not to be responsible for
the characteristic melting-point of phosphorus, it may nevertheless be reasonable, in view of what we know, *e.g.*, of other
substances and their melting-points, to *postulate* its existence.

The above is important for the following reason. If we, as a
matter of actual fact, saved the law of the melting-point of

phosphorus from falsification, by announcing that a thing which does not melt at the temperature in question is not phosphorus, then we would almost certainly *not* want to say anything decisive about the definition of phosphorus. We would rather say something like this: Perhaps the melting-point can be used for the purpose of defining phosphorus, but perhaps also there will be found some 'deeper' quality of the substance, which will explain why *phosphorus* melts at just 44° C. But irrespective of which alternative will finally be chosen, we wish at present to adhere to the norm that phosphorus melts at 44° C. Whether it is because the temperature defines phosphorus, or whether it is because it only indicates some infallible criterion of that substance, is a question not as yet considered, and one that need not be settled in this connexion.

In discussing the example we were concerned only with the criteria of phosphorus, and assumed all other properties which might enter the course of the experimental inquiry, to be actually well-defined. It is clear that this assumption is, on the whole, fictitious. The law of the melting-point was made to escape falsification by doubting the 'real' presence of phosphorus, but we might just as well have achieved the same end by doubting the 'real' absence of the assumed melting-temperature. It is hardly possible to tell in advance which way of escape is more likely to be used in practice; that the second way may also be a reasonable one is clear upon slight reflection as to the complications which may enter the measurement of temperature, such as variations in atmospheric pressure, incomplete isolation of the measured body from its surroundings, defective thermometers, *etc.*

The example discussed illustrates certain general principles. These may be summed up as follows:

i.   Ideal induction presupposes that we know whether the properties related in the conclusion are logically independent or not.—*E.g.*, induction is not ideal if hesitation as regards the definition of phosphorus can arise from failure to observe a property which is originally thought to be connected with phosphorus by law.

ii. Not even in scientific practice is this demand always satisfied.—*E.g.*, hesitation as regards the definition of phosphorus will normally not arise until scientific experimentation with phosphorus has already begun.

iii. Actual induction contributes to the satisfaction of the ideal demand and thus to the gradual formation of well-defined concepts.—*E.g.*, only after a tentative exclusion of unessentials do we get to know which features of phosphorus it is convenient to use for definition, and which for the establishment of natural laws.

iv. It is plausible, in normal cases, to think of the formation of well-defined concepts as a process which never comes to an end.—*E.g.*, we must be prepared to accept that discoveries concerning the microphysical structure of chemical substances will influence their definitions. This influence is illustrated by the discovery of isotopes.

Thus it is fictitious to take definition and induction as sharply separated activities. Separation certainly answers to an ideal demand, but if carried out in practice, it would lead to a dogmatism which is irreconcilable with the truly scientific spirit.

The fact that induction presupposes definition and definition induction does not, properly interpreted, lead to a vicious circle. There is circularity only if we think of definition and induction in their pure form of associating and dissociating characteristics, either by verbal convention, or on the basis of factual observation. In practice, however, we seldom or never have ideal definition and ideal induction, but rather definition which is tentative or anticipative with regard to what will be observed, and induction which is prepared to interpret the results of observation in the light of what will be verbally convenient. If we wish to use an image, the successive movement of definition and induction is not, therefore, to be described as a circle, but as a spiral, each branch of which is an inductive conclusion when viewed from below, and a definition when viewed from above. The progressive way of science also resembles a spiral, in that the steps through ascending levels of induction and formation of concepts are continuous and imperceptible rather than abrupt and startling.

Even if it is plausible to think that the formation of concepts never comes to an end, it does not follow that definition is a haphazard process in the course of which no general directions are discernible. It would be an important task for methodological study to investigate in detail such 'directional movements' of scientific procedure, and to estimate their relevance to the general development of human thought. One important movement has already been sketchily indicated in our above example from chemistry. When different chemical substances were originally discovered and named, the criteria used for classifying them were, in the first place, macroscopic properties of a qualitative nature such as colour, smell, weight, softness, *etc*. These properties were later found to be connected by law with other macroscopic characteristics of a quantitative nature, such as fixed melting-point at constant pressure, specific gravity, *etc*. It was thought convenient, for several reasons, to regard the presence of those quantitative criteria as more reliable and therefore more 'real' than the qualitative, and so a shift in the formation of scientific notions of chemical substances took place, from the qualitative to the quantitative in the macroscopic sphere. At an even later stage, however, it became possible to account for the presence of the macroscopic features by reference to underlying hypothetical microphysical features. Since this introduced further simplification and unity in the scientific conception of the world, it became convenient to locate 'reality' in the microphysical and to refer, say, to the positions of elementary substances in the periodic system as their most 'essential' characteristics. This passing from the qualitative to the quantitative and from the macrophysical to the microphysical, and the general tendency to withdraw scientific reality from immediate touch with the sensible world, has been of great importance. Not only has it immensely stimulated the advance of science and technique, but it has also exerted a strong and perhaps not equally happy fascination over philosophic thinking from the time of the Greeks up to the present day.

3. Induction and the Rectification of Laws

We shall here discuss another example, which is also well-known in the literature:[1]

We know, from physics as well as from everyday life, a number of laws associating impact and movement. These are, sometimes at least, inductive conclusions. We have, *e.g.*, observed that the impact of one billiard-ball upon another is followed, so far as our experience goes, by the movement of the second ball. From this we conclude that *whenever* one billiard-ball strikes another, the latter will be moved, or, as we may also express it, that impact (of one billiard-ball upon another) is a Sufficient Condition of (the second ball's) movement. This is a Universal Implication $H \subset A$, if H denotes 'impact' and A 'movement.'

What of the possibility of the law coming to be refuted by experience?

In order to answer the question, let us consider how we should react, if it actually happened one day that a billiard-ball was struck by another but was left unmoved.

It can hardly be imagined that in such a case we should instantly say that the previously enunciated law had been falsified. Instead of this we would investigate more closely the circumstances under which the impact had taken place, in order to find an 'explanation' of what had happened, *i.e.*, to show, generally speaking, that what happened was in accordance with some other law, operating against the one which we were considering in the first place. Suppose, for example, that we found that the second ball was fixed to the table and could not move at all. This would justify us in saying that the law was not false, but that the cause could not operate because of the presence of a counteracting cause, the ball being fixed to the table.

All this may seem extremely trivial. Nevertheless we can learn a great deal from it. It shows that in order to 'save' a Law of Nature from refutation, it is not necessary to declare that the terms of the law are logically dependent on one another. We need not doubt, in the example under discussion, that the

[1] Cf. *The Logical Problem of Induction*, pp. 52–5

first ball 'really' struck the second and that the second 'really' did not move. (We *may*, of course, doubt it and thus adopt the course discussed in the previous section.) For the law, as originally enunciated, was still incomplete in its formulation. Instead of saying that whenever one ball strikes another the second one will move, we intended to say that whenever one ball strikes another the second ball will move, *provided certain circumstances are present, certain conditions fulfilled.* Thus if there is impact without movement, we need not deny the truth of the law, but only the presence of valid conditions for its application. And this, surely, is what we very often do in similar cases.

The terms of the law, in other words, are more complex than was apparent from the first loose formulation of it. The complexity here enters in the form of a conditional clause for the law's validity or applicability.

We can conceive of the conditional clause as a conjunction of a number of properties. (Some of the properties may themselves be disjunctive.) Since the logical product of the properties of the clause and the impact is supposed to be a Sufficient Condition of movement, any one of the properties themselves is a Contributory Condition (p. 73) of movement. If there is impact but no movement in a certain case, then some of the Contributory Conditions of the clause must be absent. Consequently its negation must be present. This negation is a Counteracting Condition (p. 73) for the absence of movement in the presence of impact.

The clause thus prescribes the presence of a number of Contributory Conditions of movement or, which is the same, the absence of certain Counteracting Conditions. Which mode of expression, the 'positive' or the 'negative,' is actually used in laying down the clause is largely a matter of verbal convenience. We may, *e.g.*, say that impact is sufficient to produce movement, provided some Contributory Conditions are present, such as that the impact has a certain minimum force, and that the surface over which the balls move has a certain minimum smoothness, and provided some Counteracting Conditions are absent, such as that the ball to be moved is not fixed to the table, that it is not acted upon by forces of a certain kind, exceeding a maximum amount, and so on.

For practical purposes, however, reference to the clause is usually omitted in enunciating the law, or it is at most referred to in the form of a *Ceteris-Paribus*-Clause stating that impact will in future also be followed by movement, provided the new instances resemble the old ones in all relevant circumstances. Detailed formulation of the clause is thought unnecessary, either because the details are concerned with exceptional circumstances, which very seldom need be taken into account, or because the details are so trivial and self-evident that their fulfilment in 'normal' cases is taken for granted without special mention. Besides this, we have at the back of our minds the idea that although the law is 'in practice' left incomplete in its formulation, it is always 'theoretically' possible to formulate it in full, if need be.

This idea needs to be more closely examined. Let us ask the following question: How would it be possible to know that all circumstances necessary for the full formulation of the law have been taken into account? Two principal answers are possible.

The first answer consists in the enumeration of a finite number of circumstances, of which is asserted that they jointly cover the whole range of the *Ceteris-Paribus*-Clause. If, in spite of the presence of these circumstances of the clause, impact is not followed by movement, we have to admit that the inductively established law which we are discussing, is falsified.

It is not, however, certain that this answer would recommend itself as the one which is most in accordance with the actual practice of science. The enumeration of the circumstances constituting the clause might well have an air of arbitrariness which the scientist wishes to avoid.

The second answer runs as follows: Whether the enumeration of a finite number of circumstances covers the whole range of the clause will depend upon future experience concerning the production of movement by impact. If, in spite of the presence of the enumerated circumstances, impact is not followed by movement, we simply declare the enumeration of circumstances incomplete. In other words, as soon as the ball is struck but left unmoved, we say that there must still be some circumstance relevant to the applicability of the law, which has not yet been taken into account, and which is absent in this case. Thus the truth of the inductively established law is made the standard determining the content of the clause, and not conversely. The truth of the inductive conclusion serves as the norm which guides us in the search for new qualifications to be added, for the purpose of getting a complete and exhaustive formulation of the law aimed at in making the generalization from experience.

It would, to be sure, be an oversimplification to say that the second answer is in better accord with the actual practice of science than the first. The truth seems to be somewhere in the middle. Usually we are prepared at first to let ourselves be guided by the assumed truth of the law, for the purpose of adding new circumstances to the content of the *Ceteris-Paribus-*Clause. We stick to this attitude, so long at least as it does not cause us much difficulty in detecting the circumstances in which cases where the law holds differ from those in which it apparently fails to do so. Even if the detection of circumstances encounters insurmountable difficulties, it is in our power to *postulate* their presence and thus allow the law to escape invalidation. But the more such postulates are needed, and the more remote their confirmation appears, the more do we become inclined to regard the enumeration of properties in the clause as definitive and exhaustive of its real content, and to let the law be invalidated if it fails to hold in the presence of those properties.

The general relevance of the example discussed may be summed up as follows:

i. Ideal induction presupposes that the complexity of conditions is well-defined in terms of the properties of a certain set.—*E.g.*, induction is not ideal if hesitation as regards the formulation of a law connecting impact with movement can arise from failure to observe movement in spite of impact.

ii. Not even in scientific practice is this demand always satisfied.—*E.g.*, hesitation as regards the full formulation of a law connecting impact with movement will normally not arise until experimentation with moving bodies has already begun.

iii. Actual induction contributes to the satisfaction of the ideal demand and thus to the gradual rectification of defectively formulated laws.—*E.g.*, only after some failures to observe movement following upon impact do we get to know the circumstances which are relevant to the applicability of a law connecting impact with movement.

iv. It is plausible, in many cases, to think of the gradual rectification of defectively formulated laws as a process which never comes to an end. In these cases there is no falsification of laws, but only correction of them so as to answer to demands for increasing exactitude.

4. *Remarks on the Historical Development of the Logic of Inductive Truth*

In this section we shall briefly summarize the relevance of some of the best-known treatments of induction to the problems which have been discussed in the last three chapters.

The first attempt at a systematic treatment of induction was made by Aristotle. His use of the term is ambiguous; ἐπαγωγή means, for Aristotle, partly a process of inference which, from the enumeration of instances of a theory or law (in our sense of the words), concludes to the theory or law itself, and partly a process of definition which, from the observation of single instances, 'abstracts' a general notion exemplified in them. The inferential process is further understood, sometimes as a demonstrative argument following from a complete enumeration of instances, and sometimes as a non-demonstrative argument following from an incomplete enumeration. Aristotle's treatment of induction is intimately connected with the rest of his logic and his epistemology. This makes it difficult to relate it to modern discussion. It seems fair, however, to summarize Aristotle's contributions to the subject as follows:

i. Aristotle was the first to point out the non-demonstrative character of the type of inference which we treat under the name of induction, and to contrast it with conclusive reasoning. The contrast, however, was obscured by his own terminology which has become established in traditional logic.

ii. Aristotle was aware of the double aspect of inductive method as a process of inference and as a process of definition (formation of concepts).

A radical advance beyond the standpoint of Aristotle in the study of induction and scientific method was made by Francis Bacon. (There are interesting anticipations of Bacon's views on induction among the schoolmen.) He spoke of the enumerative induction of traditional logic as a *res puerilis*. The only useful way of making inductions is that which proceeds *per rejectiones et exclusiones debitas*, *i.e.*, through the elimination of concurrent possibilities. An inductive conclusion from enumerated positive instances of a law 'is no conclusion, but a conjecture' and therefore 'utterly vicious and incompetent.' Only Induction by Elimination, as he says, *necessario concludat*.

In other words: Laws of Nature are not verifiable, meaning that their truth never follows from the verification of a finite number of their instances, but they are falsifiable in the sense that their falsehood necessarily follows from the falsification of a single instance. It is the immortal merit of Bacon to have fully appreciated the importance of this asymmetry in the logical structure of laws. He stressed one of the distinguishing features of scientific method, and thereby became responsible for one of the greatest advances ever made in the history of methodology (logic of method).

Macaulay, in his famous essay on Lord Bacon, was anxious to refute the claim that Bacon had invented the inductive method. Certainly, not only the enumerative but also the more advanced eliminative mode of induction had been used long before Bacon. The rules of Induction by Elimination are constantly employed in ordinary life, whenever there is an attempt at methodical 'causal' reasoning from experience. Most of them are, in practice, familiar even to primitive man.

What Bacon accomplished was a description of some of the main features of Eliminative Induction. He was an innovator, not in the realm of method, but in the realm of the logic of method. Macaulay, therefore, was not justified in his attempt to refute Bacon's claims to novelty in logic. It is not a fair statement of historical facts to say, as Macaulay does with reference to Aristotle: 'Not only is it not true that Bacon invented the inductive method; but it is not true that he was the first person who correctly analysed that method and explained its uses.' Neither Aristotle nor Bacon completed the task of correctly describing the inferential mechanism of induction, but Bacon's independent contributions to the subject suffice to justify his claim that he was *in hac re plane protopirus, et vestigia nullius sequutus.*

I think we can assent to Macaulay's statement that the nature of Bacon's achievements in the study of induction 'is often mistaken, and was not fully understood even by himself.' Bacon's mind—in particular, it would seem, at the time when he first designed his *Instauratio Magna*—was strongly dominated by the idea that his philosophy of induction was to help man to discover new truths in science (p. 19), and further, to promote the application of scientific discoveries for technical purposes. Indeed, if there is anything more important in Bacon's work than his contributions to methodology, it is the stress he laid upon the close relation of science to technique and thereby to life and society. Bacon's work, however, as a creator of new values, and his work as an innovator in logic should be kept apart, and the fact that he was himself mistaken as to their mutual relationship should not lead us to false conclusions in estimating their separate importance.

The task of induction, according to Bacon in the *Novum Organum*, is that of finding the forms of given natures. These terms he had inherited from medieval philosophy, though he proposed to make a new use of them. We cannot here follow in detail the development of Bacon's views as to the purpose of inductive method, and the various interpretations which he put upon the relation of nature to form. The relation especially in his earlier writings, has a causal aspect, the form being sufficient for the production of the nature. On the other hand, nature and form do not differ as to their occurrence in time, but are simultaneous. The type of laws which Bacon was seeking are thus what Mill called Uniformities of Co-Existence. Later on in his writings, the relation of nature to form approaches more and more to that of secondary to primary qualities. Nature and form differ as *apparens et existens, aut exterius et interius, aut in ordine ad hominem et in ordine ad universum.* Ultimately Bacon's doctrine of natures and forms has little in common with scholastic philosophy, but much resembles leading ideas in the thought of Galileo and Descartes.

From the point of view of logic it is important to note that the form is a Necessary-and-Sufficient Condition of the nature. For Bacon says: *Etenim Forma naturae alicujus talis est ut, ea posita, natura data infallibiliter sequatur. Itaque adest perpetuo quando natura illa adest.* . . . *Eadem Forma talis est ut, ea amota, natura data infallibiliter fugiat. Itaque abest perpetuo quando natura illa abest.* The presence of the form universally implies the presence of the nature, and the absence of the form the absence of the nature. Consequently, form and nature are universally equivalent. The type of law which Bacon seeks is thus that of Universal Equivalence.

Bacon's description of the mechanism of elimination is, in substance, identical with his description of the three 'tables.' The description of the *tabula essentiae et praesentiae* corresponds to our description of the Simple Case of the Direct Method of Agreement, and the description of the *tabula declinationis sive absentiae in proximo* to our description of the Simple Case of the Inverse Method of Agreement. The *tabula graduum sive tabula comparativae* is, broadly speaking, an application of the principles of the first two tables for the purpose of detecting nomic connexions in the variations of measureable quantities (p. 83).

Taken together, Bacon's tables answer to the Simple Case of the Double Method of Agreement which, as we know, is a method for investigating Necessary-and-Sufficient Conditions and for establishing Universal Equivalences.

There are also incidental contributions to the logic of elimination in Bacon's description of the 'prerogative instances.' The first of these instances, the *instantiae solitariae*, answer approximately to a state of Perfect Analogy attained in a minimum number of steps by use of the Direct Method of Agreement in the Simple Case, and the second, the *instantiae migrantes*, to the same state in connexion with the Double Method of Agreement in the Simple Case. The latter can also be interpreted as equivalent to the Method of Difference in Mill's sense.

There is no distinct acknowledgement of Plurality or of Complexity of Conditions in Bacon's account of the mechanism of elimination. In two senses, however, forms may be said to be complex conditions. First, Bacon sometimes talks of the form as a specific difference of a proximate genus. This implies that the form, as Necessary-and-Sufficient Condition of the nature, is the logical product of two factors. Secondly, Bacon assumes the existence of a small number of simple forms, of which all the variety of the world is ultimately composed. It is not quite clear how this is to be understood; it appears that Bacon thought of forms as presence-functions of members of a finite set of (simple) forms.

The idea that there is a small number of forms at the basis of all variety can be understood as a Selection Postulate, in the inclusive form of a Postulate of Limited Variety, and thus as a supplementary premiss of induction. It is further related to another favourite idea of Bacon, *viz.*, that of a complete catalogue of human knowledge.

The other supplementary premiss of induction, the Deterministic Postulate, is never explicitly referred to by Bacon. His philosophy of induction implicitly rests upon the axiom that any nature has one form and one only.

Against the background of his implicit Deterministic Postulate, his explicit Selection Postulate, and his neglect of Plurality and Complexity of Conditions in the description of the mechanism of elimination, it is not difficult to understand how Bacon came to entertain the idea that Induction by Elimination is capable of attaining absolute certainty. He contrasts the enumerative method, which *periculo exponitur ab instantia contradictoria*, with the new method, *quae ex aliquibus generaliter concludat ita ut instantiam contradictoriam inveniri non posse demonstretur.* The exclusion of falsehood will thus necessarily lead to true inductive conclusions.

Bacon gradually became more and more aware of the difficulties which the demand for well-defined notions imposes on an ideal use of induction. In the first part of the *Novum Organum* he condemns all commonly received scientific conceptions as worthless. Until their status has been revised, use of the eliminative method will remain in some degree imperfect. For the method required for the formation of well-defined scientific notions, he also uses the name induction; it is, indeed, noteworthy that the first time induction is mentioned in the *Novum Organum* it refers, not to the establishment of laws, but to the definition of concepts.

It seems that it was Bacon's original intention to deal with the formation of concepts in the first part of the *Novum Organum* and with the logic of elimination, or the establishment of laws, in the second, but that this plan was not systematically carried out because he became somehow confused over the way in which definition and induction are related. The two processes are so intertwined that they cannot be performed altogether separately, and yet ideal use of induction requires that the process of definition shall have come to an end. Bacon ultimately conceives of a successive use of them: laws must first be established on the basis of commonly received notions, and then the notions must be rectified so as to make the rectification of the laws possible. There is, however, no clear description of the *fortiora auxilia in usum intellectus* needed for the rectification of conceptions and laws.

The difficulties over definition seem to imply that the idea of limited variety was to be given up, or at least modified. Thereby the original association of Eliminative Induction and absolute certainty also becomes debatable. On the whole, Bacon grew, in the progress of his work, increasingly sceptical and critical as to the power and perfection of his inductive method. At the same time his insight into the nature of induction was growing increasingly profound.

The literature on Bacon is extensive and opinions on his work vary a great deal. Lalande[1] says: '*Bacon est le plus discuté des philosophes illustres.*' It seems to me that the best account of Bacon's contributions to the Logic of Induction is found in Robert Leslie Ellis's *General Preface* to the Ellis-Spedding edition of Bacon's *Philosophical Works*. Here should be mentioned also the illuminating papers by Broad on *The Philosophy of Francis Bacon* (1926) and Kotarbiński on *The Methodology of Francis Bacon* in Studia Philosophica *1* (1935).

[1] *Le Problème de l'Induction* (1928).

Wholly in the spirit of Bacon is Robert Hooke's posthumous work *A General Scheme or Idea of the Present State of Natural Philosophy* published in 1705. The influence of Bacon is also strongly present in the methodological ideas of Boyle and Newton.

Bacon's idea of an 'alphabet' of simple forms and Hooke's conception of the eliminative method as a 'philosophic algebra' may be regarded as a counterpart, in the inductive philosophy of England, to the idea of a *Mathesis Universalis* in the deductive philosophy of continental Europe at the same period.

In an unfinished article under the title *Hypothèse* for the great *Encyclopédie*, the physicist Lesage gave a noteworthy description of the logic of elimination in the spirit of Bacon's tables. The article was written about 1750, but was not published until 1813. Lesage paid special attention to the question how the initial set of concurrent possibilities may be limited so as to make useful elimination practicable. Due acknowledgement of the eliminative method is also found in the discussion of methodological topics in Dugald Stewart's *Elements of the Philosophy of the Human Mind* (1792, 1813, 1827).

Next in importance to the contributions of Bacon to the Logic of Inductive Truth, come the contributions of John Stuart Mill. Mill's *Logic* appeared in the year 1843. A good many of Mill's ideas had, however, been already anticipated in J. F. W. Herschel's *A Preliminary Discourse on the Study of Natural Philosophy* of 1830.

Mill defines induction as 'the operation of discovering and proving general propositions.' This is not very clear; it suggests to us that Mill, like Bacon, confused the two fundamental aspects of discovery and logical analysis. He explicitly talks of the inductive methods as 'methods of discovery' as well as 'methods of proof.' Still it must be admitted that Mill, on the whole, kept the two aspects apart. For the logical study of induction he rightly set no other goal than that of systematizing the ideas 'conformed to by accurate thinkers in their scientific inquiries.'

To Bacon in the *Novum Organum*, the methodical use of induction was intended to connect external features of bodies with structural properties by Universal Equivalence. Mill, following Lesage and Dugald Stewart, regarded the foremost task of inductive inquiry as one of tracing the causal connexions in the succession of events. This task has a double aspect: to find the effects of given causes, and the causes of given effects.

The logical relation of nature and form in the philosophy of Bacon is much clearer than the relation of cause and effect in that of Mill. Mill gives two definitions of the word 'cause' which are not logically identical. 'The cause,' he says, 'is the sum total of the conditions . . . which being realized, the consequent invariably follows.' This must imply that the cause is a Sufficient Condition of the effect. But he also says that 'the invariable antecedent is termed the cause.' This strongly suggests that the cause is a Necessary Condition. On the other hand, he recognizes that a phenomenon may have several distinct causes, which clearly shows that, unlike Lesage, he cannot, in general, be thinking of the cause as a Necessary-*and*-Sufficient Condition.

The matter becomes still more confused, when Mill later on refuses to let causal relationship mean invariable succession in time, and adds that the succession must be invariable and *unconditional*. Cause is now defined, both as that on which the effect is invariably and unconditionally consequent, implying that the cause is a Sufficient Condition, and as the effect's invariable and unconditional antecedent, implying that the cause is a Necessary Condition. What Mill meant by 'unconditional' is difficult to discover. Actually, he employs the term for different purposes, but the use which he makes of it in connexion with his description of the four experimental methods, strongly indicates that the 'real' meaning which he attached to 'cause' implies it to be a Sufficient Condition, and that the resort which he makes to the word 'unconditional' is a somewhat confused way of finding an escape from difficulties arising from the fact that he also, at times, uses 'cause' to mean Necessary Condition.

As is to be expected from the above, Mill's description of the mechanism of elimination is to some extent marred by his failure clearly to see and to state the logical relations of causally connected terms.

Mill's Method of Agreement corresponds roughly to what

is called in this book the Simple Case of the Direct Method of Agreement. On the hypothesis that 'cause' with Mill implies Sufficient Condition, it would follow that his Method of Agreement can be properly used for ascertaining the effects of given causes, but *not* for the converse task of finding the causes of given effects. This, however, was not clear to Mill. After having given a not altogether correct description of the eliminative mechanism of the method in application to the first task, he continues:

'In a similar manner we may inquire into the cause of a given effect. Let *a* be the effect . . . . If we can observe *a* in two different combinations, *abc* and *ade*; and if we know, or can discover, that the antecedent circumstances in these cases respectively were *ABC* and *ADE*; we may conclude *by a reasoning similar to that in the preceding example*,[1] that *A* is the antecedent connected with the consequent *a* by a law of causation. *B* and *C*, we may say, cannot be causes of *a*, since in its second occurrence they were not present; nor are *D* and *E*, for they were not present on its first occurrence. *A*, alone of the five circumstances, was found among the antecedents of *a* in both instances.'

Here Mill has obviously failed to see that the fact that the two instances agree in *A* but differ in *B*, *C*, *D*, and *E* has no bearing whatever upon the question of finding, by means of elimination, a cause in the sense of a Sufficient Condition of *a*. Actually, *no* characteristic has been eliminated as a possible Sufficient Condition of *a*; all we have done is to produce two confirming instances of the Universal Implication $A \subset a$ and one confirming instance of each of the Universal Implications $B \subset a$, $C \subset a$, $D \subset a$, and $E \subset a$. Thus Mill's Method of Agreement, when applied to the task of finding Sufficient Conditions, is no method of elimination at all, but merely Induction by Enumeration.

This was not clearly grasped by Mill, though he felt there was something defective about his method. As already hinted at, he made an unsuccessful effort to account for this defect by declaring that the cause should be, not only an invariable, but also an 'unconditional' antecedent, and by bringing in the

[1] Italics mine. The passage is in bk. iii, ch. viii, § 1

possible Plurality of Causes. All this is beside the point; it should be observed that if 'unconditional invariable antecedent' is interpreted, as is most natural, as a kind of Necessary Condition, then Mill's Method of Agreement actually *has* the power of proving—in combination with due supplementary premises—that one factor is such an antecedent in relation to a given factor.

On the same hypothesis, that 'cause' with Mill implies Sufficient Condition, it further follows that his Method of Difference can be properly employed only for ascertaining the causes of given effects, but not the effects of given causes. On this point also there is much obscurity in Mill's description.

It is a peculiarity of Mill's account of the Method of Difference that it is exclusively concerned with the limiting case— corresponding to Bacon's *instantiae migrantes*—when one positive and one negative instance of the phenomenon in question are sufficient to establish a Perfect Analogy. He did not clearly realize that there is a form of the Method of Agreement— corresponding to Bacon's *instantiae solitariae*—which is equally powerful, though in relation to a different purpose. This failure of his is connected with his confused recognition of the defectiveness of the Method of Agreement when used for the purpose of finding causes, *i.e.*, Sufficient Conditions. In this way Mill came to believe that the Method of Difference was somehow *eo ipso* more efficient as a method of elimination than the Method of Agreement. His charge against the latter method, however, becomes unjustified once the logical mechanism of the two methods is made quite clear.

In Mill's account of his Joint Method too there are several perplexities. Some of them are connected with ambiguities in the terms 'cause' and 'effect,' others with the failure to separate sharply the two principal forms of the Joint Method which we have called in this book the Double Method of Agreement and the Joint Method of Agreement and Difference respectively.[1]

Mill's Method of Residues is, strictly speaking, no method of elimination at all. Its use, certainly, answers to familiar procedures in science, but reasoning based upon it is extremely weak unless supported by powerful supplementary premises.

[1] Cf. Jackson, *Mill's Joint Method I–II* in Mind *46–7* (1937–8).

If, to use Mill's own example, we have the antecedents *ABC* followed by the consequents *abc*, and from this, in combination with the assumed knowledge that *A* and *a*, and *B* and *b* are causally connected, infer that *C* is the cause of *c*, then the argument, to possess demonstrative value, must *inter alia* exclude the possibility that one and the same cause may have several effects, or that a factor may have several Necessary Conditions.

Of Mill's Method of Concomitant Variation as a canon of elimination we can say the same as of Bacon's *tabula graduum sive tabula comparativae*.[1]

There is explicit acknowledgement both of Plurality and of Complexity of Conditions in Mill's philosophy of induction. The second phenomenon appears in his description of the eliminative mechanism only in connexion with the Method of Difference, when Mill says that the sole factor, besides the conditioned factor, in which the two instances differ is the cause or 'an indispensable part of the cause.' The 'indispensable part of the cause' means, in our terminology, Contributory Condition (not necessarily Indispensable Contributory Condition). A complex cause is thus with Mill a product[2] or conjunction of properties.

There is an interesting passage in which he also mentions complex Necessary Conditions which, as we know, are sums or disjunctions of properties. Mill, however, mistakenly talks of this case as an instance of Plurality and not of Complexity of Causes. The passage in question runs:

'The plurality may come to light in the course of collating a number of instances, when we attempt to find some circumstance in which they all agree, and fail in doing so. We find it impossible to trace, in all the cases in which the effect is met with, any common circumstance. We find that we can eliminate *all* the antecedents; that no one of them is present in all the instances, no one of them indispensable to the effect. On closer scrutiny, however, it appears that though no one is always present, one or other of several always is.'

[1] Cf. above p. 83 and p. 154. Cf. also Nicod, *Le problème logique de l'induction* (1924), p. 24.
[2] Mill, incidentally, talks of 'sums' of antecedents meaning, in our terminology, logical products.

Here the antecedents which are eliminated (cf. above p. 93) must be initially possible simple Necessary Conditions. The alternatively present antecedents are possible Substitutable Requirements and their sum a possible complex Necessary Condition. The use of the word 'cause' to mean Substitutable Requirement is not uncommon, both in ordinary discourse and in science. It is, by the way, a principal use of the word in connexion with traditional doctrines on so-called Probability of Causes. (Cf. Chap. X, §§2 and 5.)

Following Archbishop Whateley, Mill regarded inductive inference as a sort of demonstrative reasoning from partly suppressed premisses. Therefore, he says, 'the business of Inductive Logic is to provide rules and models (such as the Syllogism and its rules are for ratiocination) to which if inductive arguments conform, those arguments are conclusive, and not otherwise.' (Cf. above p. 129.)

The supplementary premiss which Mill has in mind is a kind of Deterministic Postulate, called by him the Law of Universal Causation. As his standard formulation of it we may take the following: 'Every event, or the beginning of every phenomenon, must have some cause; some antecedent, on the existence of which it is invariably and unconditionally consequent.' It is obvious that 'cause' here must imply Sufficient Condition. Though Mill does not sharply distinguish between the causes of an event, as such, and the causes of an event which are present in an individual case of the event's occurrence, it is also fairly obvious from the above that Mill conceived of his Law of Universal Causation as implying that every event, in every instance of its occurrence, is accompanied by a cause, *i.e.*, one of its Sufficient Conditions, and thus is, to use our terminology, determined. (Cf. above p. 72 and p. 75 f.)

Mill's idea that the Law of Universal Causation was itself an inductive conclusion and yet capable of some sort of 'proof' is extremely obscure and has been frequently criticized. Much of the confusion, I think, arises from the fact that Mill believed the content of the supplementary premiss to be the same for all inductive inferences. As we have seen (p. 132), there is a weakest form of the Deterministic Postulate, corresponding to Mill's Law of Universal Causation, which satisfies this condition and which, moreover, cannot be refuted by experience. Of it we can truly say with Mill: 'To the law of causation—we not only do not know of any exception, but the exceptions which limit or apparently invalidate the special laws, are so far from contradicting the universal one, that they confirm it; since in all cases which are sufficiently open to our observation, we are able to trace the difference of result, either to the absence of a cause which had been present in ordinary cases, or to the presence of one which had been absent.' But we have also seen (p. 132 ff.), that *this* supplementary premiss, 'being thus certain,' is not the same as that much stronger premiss which would be 'capable of imparting its certainty to all other inductive propositions which can be deduced from it.'

The necessity of adopting a Selection Postulate as a second supplementary premiss of induction was never explicitly avowed by Mill. But, just as the Deterministic Postulate may be said to be implicitly present in Bacon's writings, so in a similar manner the Selection Postulate, in the exclusive rather than the inclusive form, may be said to underlie Mill's mode of reasoning.[1]

Mill was not puzzled to the same extent as Bacon by the fact that the use of induction contributes, not only to the establishment of Laws of Nature, but also to the formation of concepts. Under the names of Composition of Causes and Intermixture of Effects he collected a number of interesting though not very systematic observations on the way in which induction contributes to the rectification of laws.

Even if we are not prepared to assent to Jevons's well-known contention that 'Mill's mind was essentially illogical,' we have to admit that there is very much obscurity and an outstanding number of downright errors of thought in Mill's treatment of induction. I cannot avoid the impression that Mill, as a logician, was much inferior to Bacon, however fantastic and incoherent the views of the latter may appear to the modern reader.

However, the obvious weaknesses of Mill's system should not lead us to underrate its importance. Not only is Mill's treatment of induction full of valuable suggestions and acute observations on many points of detail. Most of its main ideas

[1] Cf. Keynes, *op. cit.*, p. 271.

are substantially sound. In the history of the subject here called the Logic of Inductive Truth, he is the last great innovator and constructive mind on a large scale.

An exposition of Mill's methods, usually in combination with some criticism, has become traditional in textbooks on induction and methodology. A good systematic and critical monograph on Mill's Logic of Induction still remains to be written.[1]

A contemporary of Mill is William Whewell. His enterprise in the philosophy of induction was undertaken in the grand manner and bears some resemblance to the intentions of Francis Bacon, whose true successor he considered himself. He was, however, more interested in the problems of scientific discovery than in the logical analysis and reconstruction of arguments, a subject of which he had little understanding and to which he did not make any fresh contributions. He had a vivid sense of the importance of induction for the formation of concepts (and *vice versa*), but his opinions on this topic are rather obscure, owing, it seems, to the unhappy influence of Kant's philosophy. (On Whewell cf. above p. 19 and 23 f.)

Very unsystematic, but rich in suggestive ideas in the realm of methodology, is Jevons's *Principles of Science* (1874). Jevons began, but never completed, an examination of Mill's philosophy, part of which was published in the Contemporary Review under the title *John Stuart Mill's Philosophy Tested*.[2] Jevons's chief merit in the field of induction consists, I think, in the very strong emphasis which he laid on the hypothetical, *i.e.* non-demonstrative character of genuine inductive reasoning. In Jevons's time insight into the non-demonstrative nature of induction was largely obscured, partly owing to the influence of Mill,[3] and partly owing to the trends then current in the philosophy of probability.

[1] No ultimate estimation of Mill's views on induction is possible unless attention is also paid to the, sometimes peculiar, opinions which Mill held as to the nature of deduction. There is a good monograph on Mill's Logic of Deduction by Jackson: *An Examination of the Deductive Logic of J. S. Mill* (1941).

[2] Reprinted posthumously together with a note on the Method of Difference in *Pure Logic and other Minor Works* (1890).

[3] Cf. the historically interesting polemic against Jevons's views which is to be found in the preface to the 3rd edition of Th. Fowler's *The Elements of Inductive Logic* (1876).

A clear insight into the nature of induction and its relation to hypothesis and demonstration is also found in the writings of Peirce. His are, in fact, some of the best accounts of the broader philosophic and methodological aspects of induction which exist. To the formal logic of inductive methods, however, Peirce did not directly contribute.

Important contributions are to be found in the section on induction and analogy in J. M. Keynes's *A Treatise on Probability* (1921) and in Jean Nicod's thesis *Le problème logique de l'induction* (1923). Keynes appears to be the first to have distinctly recognized that inductive inference, in order to become conclusive, must be supplemented with *two* premisses, corresponding to our Deterministic Postulate and Selection Postulate respectively. His attitude as regards the Deterministic Postulate is, however, unclear. He substitutes for it, in cases of Perfect Analogy, our Postulate of Spatio-Temporal Irrelevance, which he calls the law of the Uniformity of Nature. The obscurities on this point seem to be connected with a peculiar view of identity. Nicod also is aware of the need for two supplementary premisses of induction. He makes the important observation that the Postulate of Limited Variety, in order to serve its purpose, must go further than merely to assert the finitude of the respective ranges of initially possible conditioning properties. (Cf. above p. 137.)

The importance of the work of Keynes and Nicod, however, is not so much in the field of what is here called the Logic of Inductive Truth, as in the complementary branch called the Logic of Inductive Probability. We shall therefore have occasion to return to their opinions later. (Cf. Chap. X, §5.)

Already in Nicod's book there is an approach to the problem in terms of the Logic of Conditions. It is, however, not very systematic. The first to deal systematically with the Logic of Inductive Truth on the basis of the Logic of Conditions was, so far as I know, C. D. Broad, in the first part of his important article *The Principles of Demonstrative Induction* in Mind *39* (1930). (Cf. above p. 77.) There is in Broad's paper a substantially correct restatement of Mill's Method of Agreement, Method of Difference and Joint Method, in terms of conditions.

Here also should be mentioned Karl Popper's book *Die*

Logik der Forschung (1935). Popper underlines the asymmetry of Laws of Nature in respect of decidability and the importance of falsification (*i.e.*, elimination) to the study of scientific method. Popper, moreover, is one of the first authors on these topics, who has clearly apprehended the idleness of the attempt to 'support' elimination by reference to some general presuppositions or supplementary premisses of induction.

Chapter Seven

THE LOGIC OF PROBABILITY

1. The Concepts of Probability

'PROBABILITY' is a word which is used in a multitude of very different contexts. It is not to be expected that in all these contexts it should have one and the same meaning or stand for one and the same concept. It is plausible to think that there are several concepts of probability, and accordingly several ways of abstracting, from the use of the word in discourse, an object for treatment with the instruments of logic and mathematics.

Aristotle says in the *Rhetoric* that the probable is that which usually happens. This can be regarded as the first attempt at an analysis of the word. Aristotle might be called the initiator of the so-called frequency view which, roughly speaking, sees the meaning of an event's probability in the relative frequency of its occurrence.

With the mathematical treatment of probability by Fermat and Pascal in the seventeenth century, to be followed by the Bernoullis and Laplace and others in the eighteenth, there originated another analysis of the word, which we shall call the possibility view. According to it, an event's probability is, roughly speaking, the ratio of a number of alternative possibilities 'favourable' to the event and a number of alternative possibilities as such. *E.g.*, in tossing with a coin there are two possibilities, one of which is favourable to the toss being 'heads.' Hence the probability of getting 'heads' in tossing would be 1 : 2.

The possibility view of probability was in origin more or less loosely connected with the additional demand that the possibilities in question should all be 'equal.' It is *rational* to think that the probability of 'heads' in the next toss with a certain coin is 1 : 2, only if in this particular situation, the two alternatives 'heads' and 'tails' are *equally* possible. This we normally think them to be if the coin is homogeneous, but not if it is biased.

The additional demand of equipossibility obviously raises a problem. It should, however, be observed that from the point of view of the mathematical treatment of probability, it is not necessary to include the demand of equality in the definition of the concept as a ratio of possibilities. A purified and generalized form of the possibility view was suggested by Bolzano in his *Wissenschaftslehre* of the year 1837, and renewed by Wittgenstein in the *Tractatus logico-philosophicus* (1921–22). The Bolzano-Wittgenstein definition does not mention equal possibilities. Yet it can be shown that this definition is a sufficient basis for the deductive development of the branch of mathematics known as the ('classical') Calculus of Probability.

It was, in part, the difficulties over equal possibility which, in the nineteenth century, were responsible for a revival of the frequency view. In the early forties there was an attack on the possibility view by Leslie Ellis, in his important paper *On the Foundations of the Theory of Probabilities* in the Transactions of the Cambridge Philosophical Society, by John Stuart Mill in his *Logic*, and by Cournot in his *Exposition de la théorie des chances*. Somewhat later (1866) Venn, in his *Logic of Chance*, made a systematic attempt to show how the calculus could be developed on the basis of a definition of probability as a relative frequency of an event within a sequence.

The frequency view was given a decisive impetus by von Mises in his paper *Grundlagen der Wahrscheinlichkeitsrechnung* in the Mathematische Zeitschrift of the year 1919, and in his book *Wahrscheinlichkeit, Statistik und Wahrheit* (1928, 2nd Ed. 1936).

The work of von Mises, however, also raised a serious philosophic problem which may be described roughly as follows: Just as, on the possibility view, we should not consider it rational to think that the probability of 'heads' in the next toss with a certain coin is 1 : 2, merely because there are two possibilities, one of which is favourable to 'heads,' but would demand that the two possibilities should be *equal*, so on the frequency view, we

should not consider it rational to think that the probability is
1 : 2 merely because this is the relative frequency of 'heads,'
but would demand that the distribution of 'heads' and 'tails'
over the tosses should be *random*. For, *e.g.*, if exactly every second
toss in the past had been 'heads' and every second 'tails' and if
the last toss with the coin were known to have been 'heads,'
then we should not think that the probability of 'heads' in the
next toss is 1 : 2, but rather that it is o. Von Mises tried to
embody this demand for randomness in his frequency definition
of probability, in the form of an Axiom of Irregularity.

The idea of random distribution or irregularity we shall discuss later. All that need be said here is that just as the possibility
view can be worked out independently of the additional demand
for equipossibility, so the frequency definition can be shown to
be a sufficient basis for the deductive development of the
('classical') Calculus of Probability independently of the
additional demand for randomness in the distribution of events.

Besides the frequency view and the possibility view there is
also a third traditional approach to probability. We shall call it
the psychological view. According to it an event's probability
is, roughly speaking, a measure of our degree of belief in the
event's occurrence.

This view is notoriously obscure. The interpretation of it
which will be suggested, pretends neither to be historically
adequate nor to be of practical importance. I think, however,
that the interpretation is illuminating as a sort of *Gedanken-
experiment* showing what a psychological view of probability
could be.

How are degrees of belief to be measured? There seem to be
two principal alternatives.

The first would be to identify degree of belief with something
which might be called *intensity of belief feeling*. The intensities
might be estimated introspectively and perhaps also 'extrospectively' on the basis of some physiological (*e.g.* vasomotor)
equivalent to them. This way of measuring belief does not,
however, look very promising, if only for the reason that our
belief in a great many things which we take for granted seems
to be accompanied by practically no feeling at all.[1]

[1] Cf. Ramsey, *The Foundations of Mathematics and other Essays* (1930), p. 169.

The second possibility would be to view belief, not as a feeling, but as a *disposition to act*. There is a 'classical' way of testing the strength of such dispositions, *viz.*, by proposing a bet and observing which are the lowest odds accepted.[1] This way of measuring degrees of belief is obviously much more promising than the first. Whether it is altogether satisfactory need not concern us here.

It will be sufficient for our purposes to pursue the following discussion *as though* there existed a satisfactory way of measuring partial belief. Of this way we shall assume only that it is independent of the concept of probability, *i.e.*, that measuring a degree of belief in an event does not presuppose that we have first measured the probability of the event. It is questionable whether such a condition of independence could be fulfilled, but this is also a point which need not concern us here.

As far as I know, it has never been seriously suggested that the probability of an event is *merely* a degree of belief in the event's occurrence. Since different persons may believe in the same event to a different degree, the psychological interpretation would make probability altogether 'subjective.' What, has, however, been seriously suggested is that the probability of an event is the *rational* degree of belief in it. Probability, loosely speaking, is not so much how we actually believe as how we ought to believe.[2]

Thus each of the three main traditional views of probability: the frequency view, the possibility view and the psychological view, present a dual aspect. On the first, probability is defined either as merely a frequency in a sequence or as a frequency in a random sequence. On the second, probability is defined either as merely a ratio of possibilities or as a ratio of equal possibilities. On the third, probability is defined either as merely a degree of belief or as a degree of belief which it is rational to entertain.

---

[1] Cf. Ramsey, *op. cit.*, p. 172.
[2] Cf. the following quotation from de Morgan, *Formal Logic* (1847), p. 172: 'By degree of probability we . . . ought to mean degree of belief . . . I . . . consider the word (*sc.* "probability") as meaning the state of the mind with respect to an assertion . . . on which absolute knowledge does not exist. "It is more probable than improbable" means . . . "I believe that it will happen more than I believe that it will not happen." Or rather "I *ought* to believe, etc." ' This quotation is a good illustration of the confusion between a factual and a normative aspect which is traditionally characteristic of the psychological view of probability.

On the two first views, moreover, randomness of distribution and equality of possibilities respectively seem to serve as criteria of the rationality (adequacy) of probability estimations.

In addition to the three traditional views of probability there is also a fourth approach to the analysis of the word. We shall call it the axiomatic view of probability. Roughly speaking, it takes the concept of probability as a basic notion of a deductive system which is not defined in the sense of being made explicit in terms of other concepts. All that is required is that it should obey certain axioms. The axioms are sometimes said to constitute an 'implicit definition' of probability.

Keynes' book *A Treatise on Probability* of the year 1921 might be considered as a first large scale attempt at an axiomatic treatment of probability. It dates, however, from a time when general ideas on axiomatic systems and formalizations were much less developed than they are to-day. It is for that reason mainly of historical interest. It is, moreover, not purely axiomatic in our sense, since it is biased in favour of some form of the psychological view of probability. Still, much inspiration can also be drawn from its study for modern research.

A more rigorous and philosophically neutral exposition of the axiomatic view was given by Reichenbach in the paper *Axiomatik der Wahrscheinlichkeitsrechnung* in the Mathematische Zeitschrift of the year 1932; further developed in the larger work *Wahrscheinlichkeitslehre* (1935). Here also should be mentioned the axiomatic treatment by Kolmogorov, *Grundbegriffe der Wahrscheinlichkeitsrechnung* in Ergebnisse der Mathematik und ihrer Grenzgebiete II, 3 (1933).[1]

In this chapter we shall follow the axiomatic approach to probability. In a concluding paragraph, however, we shall discuss the relation of axiomatic probability to the probability concepts of the frequency, the possibility and the psychological views, respectively. We shall try to make precise a sense in which these three traditional views in their 'bare' forms, *i.e.*, without use of the additional ideas of random distribution, equipossibility, and rationality of belief, represent three different

---

[1] An axiomatic treatment of probability was attempted by me in the paper *Ueber Wahrscheinlichkeit, eine logische und philosophische Untersuchung* in Acta Societatis Scientiarum Fennicae, Nova Series A III 11 (1945), of which the treatment of probability in this book is a further elaboration.

interpretations of axiomatic probability. The significance of the ideas of random distribution, equipossibility, and rational belief will themselves be discussed in the next chapter.

*　　*　　*　　*　　*

Axiomatic probability can conveniently be described as an 'abstraction.' The concept will be given shape in a number of declarative statements as regards its nature.

We conceive of probability as a magnitude attributed to propositions. This, however, is only a provisional formulation subject to further clarification.

On what, in the case of a given proposition, does this magnitude depend?

We shall take the view that it depends on some other propositions. We shall call these other propositions the evidence for the probability of the given proposition.

One might suggest that for the purpose of developing a logic of probability, propositions could be treated as unanalysed wholes. This is the course followed by Keynes and Jeffreys.

We shall here take a narrower view of probability. In dealing with the concept we shall presuppose a certain analysis of propositions.

By the probability of something we shall understand the probability that a certain thing has a certain property. (Or that a certain ordered set of things has the property of being a positive order of a certain relation. Cf. above, p. 48.)

By the evidence for the probability that a certain thing has a certain property we shall understand a proposition to the effect that the *same* thing has a certain (other) property.

By the magnitude or value of a probability we shall understand a real number.

It has been a matter of dispute whether the magnitude of probability needs to be a numerical magnitude or not. It frequently happens that we attribute a high, a moderate or a low probability to something, without being able to tell the numerical value of this magnitude. We frequently also compare probabilities in respect of greater and less, without specifying their difference in numerical terms. Estimates of the above kind appear, moreover, to be the only ones possible when we judge the probability of Laws of Nature. The question then arises whether the magnitudes thus spoken of are merely probabilities, the numerical values of which exist but are unknown, or whether they are non-numerical probabilities in a 'real' sense.

Keynes assumed the existence of non-numerical probabilities as distinguished from unknown numerical ones.[1] It is obviously possible to develop a logic of probability also for non-numerical magnitudes.[2] Usually, however, probabilities are unhesitatingly treated as numerical quantities. Here we shall follow this narrower course.

Given a first and a second property from the same universe, and a real number, it is either true or false that the number is the probability that a random thing has the first property on the evidence that it has the second property.

Thus any two properties from the same universe, and a real number, constitute either a positive or a negative order of a certain relation. This three-termed and non-homogeneous relation we call the probability-relation. As its symbol we use P.

The first property we shall call the conjectured property. The second property we shall call the evidence (property) or, sometimes, the field of measurement.

It is not implied that, given a first and a second property from the same universe, there will always exist a real number which indicates the probability that a thing has the first property on the evidence that the same thing has the second property. It is not necessary, in other words, that there should always be a probability that a thing has a certain property on the evidence that it has a certain other property.

The relation denoted by P is our concept of axiomatic probability.

2. *The Calculus of Probability*

We shall now outline an axiomatic and formalized deductive system for the relation P. This system will here be called the Calculus of Probability.

[1] *Op. cit.*, Chap. III, especially p. 34.
[2] Cf. B.O. Koopman, *The Axioms and Algebra of Intuitive Probability* in Annals of Mathematics *41* (1940) and *Intuitive Probabilities and Sequences* in Annals of Mathematics *42* (1941).

That the deductive system is axiomatic means, approximately speaking, that a finite number of sentences (propositions) are laid down from which all other sentences (propositions) of the system follow. The sentences (propositions) of the first group are called axioms, and the sentences (propositions) of the second group theorems.

That the deductive system is also formalized means that a finite number of rules are laid down for the derivation of theorems from the axioms.

The sentences expressing axioms and theorems are also called formulae of the calculus.

The propositions expressed by the formulae are also called principles of the calculus.

It will be convenient to use the terms 'axiom' and 'theorem,' sometimes for a formula (sentence) and sometimes for a principle (proposition) of the calculus.

The formulae of the calculus are molecular complexes of four kinds of sentences, *viz.*

i. Quantified Sentences belonging to the Logic of Properties. These sentences state that a property exists, that two properties are mutually exclusive, that one property is included in another, or that two properties are co-extensive.

ii. Sentences concerning sequences. These sentences, which occur mainly at a higher stage of the calculus, state that a sequence of properties or of numbers approaches a limit, or that the relations *Exc* or *Inc* (cf. above p. 60) subsist in sets of index-numbers of properties.

iii. Sentences stating that the relation *P* subsists in (appropriate) sets of things. Sentences of this kind will be called probability-expressions, and the propositions which they express will be called probability-propositions.

If, in a probability-expression, the name of the conjectured property is atomic, we speak of a simple probability. If it is a negation-name, we speak of a complementary, if it is a conjunction-name, of a conjunctive, and if it is a disjunction-name, of a disjunctive, probability.

    iv. Sentences of arithmetic. These will throughout be equalities (identities) or inequalities between atomic or molecular numerals for real numbers.

*   *   *   *   *

The construction of the calculus is pursued in three stages. The deductive system reached in the two first stages we call the Elementary Calculus of Probability. It embraces what is commonly known as arithmetical or discrete probability. The deductive system of the third stage we call the Higher Calculus of Probability. It deals with so-called geometrical or continuous probability.

It is characteristic of the first stage that the probability-expression contains only atomic names of properties or molecular names with a given finite number of constituents (here not more than two). Within this framework all the axioms can be stated, and some fundamental theorems, *e.g.* the Multiplication, Addition, and Inverse Principles, can be proved.

It is characteristic of the second stage that the probability-expression contains also names of properties which are presence-functions of an undetermined number n of constituents. The theorems of the first stage are generalized so as to become valid for any value of n. The generalization takes place by means of reasoning from n to $n+1$. Within this new framework we can prove, *e.g.*, some of the so-called Principles of Great Numbers, which state the convergence of a probability towards a limit.

Finally, it is characteristic of the third stage that the probability-expression contains also names of properties which are presence-functions of a non-denumerable number of constituents. The theorems on arithmetical probability of the Elementary Calculus are now generalized so as to become theorems on geometrical probability of the Higher Calculus of Probability. To give an account of the exact nature of this generalizing step—which from the point of view of logic means the transition from presence-functions of a denumerable, to those of a non-denumerable number of properties, and from

175

the point of view of mathematics from sums to integrals—is beyond the aim and purpose of the present inquiry.

Geometrical probability or the Higher Calculus will, therefore, not be treated here.

3. The Axioms

The axioms are the following six:

A1. $(E\,H)\rightarrow(P(A, H, p)\,\&\,P(A, H, q)\rightarrow p=q)$

A2. $(E\,H)\rightarrow(P(A, H, p)\rightarrow p\geqslant 0)$

A3. $(E\,H)\rightarrow((H\subset A)\rightarrow P(A, H, 1))$

A4. $(E\,H)\rightarrow(P(A, H, p)\rightarrow P(\overline{A}, H, 1-p))$

A5. $(E\,H\&A)\rightarrow(P(A, H, p)\,\&\,P(B, H\&A, q)\rightarrow P(A\&B, H, pq))$

A6. $(E\,H)((n)P(A_n, H, p_n)\,\&\,P(A, H, p)\,\&\,lim(A_n, H)\rightarrow lim(p_n, p))$

As will be seen, it is a common feature of all the axioms that they assume the respective evidence properties to exist, *i.e.*, not to be empty.

This feature is quite natural from the point of view of the 'content' of probability. In some interpretations of the calculus, to be discussed later, any probability which is relative to empty evidence will have the value 0 : 0, *i.e.*, be numerically undetermined.

The qualification of existence is more restrictive than the qualification of not being contradictory, which was used by Keynes. Any contradictory property is empty, but not all empty properties are contradictory.

A1 is called the Axiom of Uniqueness. According to it, if there is a probability in favour of the occurrence of a property in a thing on given evidence, then this probability, the evidence remaining unchanged, will be one real number and one only.

A2 is called the Minimum Axiom. According to it, negative real numbers are excluded from being probabilities.

A3 is called the Inclusion Axiom. It states that if one property is included in another, then the probability of the occurrence of the second property in a thing, relative to the occurrence of the first property as evidence, is 1.

A4 is called the Addition Axiom. According to it, the sum

of a probability and the corresponding complementary probability is 1.

A5 is called the Multiplication Axiom. It gives a rule for computing a conjunctive probability on the basis of two simple probabilities.

A6 is called the Axiom of Continuity. It states that if a sequence of properties approaches a certain property as its limit (p. 54f.), then the probabilities of the occurrences of the former property approach the probability of the occurrence of the latter property as their limit.

It should be observed that *A4* in combination with the other axioms is sufficient for the proof of the stronger proposition *T9* or the Special Addition Theorem, which is usually included among the axioms of the calculus. This simplification, as compared with the systems of, *e.g.*, Reichenbach and Kolmogorov, is decidedly in the interest of the logical economy of the deductive system.

A6 or the Axiom of Continuity seems to me to occupy a peculiar and, as it were, subordinate position among the axioms. We shall not need it until Chap. IX when we deal with Inductive Probability.

4. The Process of Inference

In the proofs there occur deductive steps of the following three kinds:

i.   Sometimes the deductive step has the form of an assertion that certain sentences entail a further sentence. This should always be understood as meaning that the conjunction of the former sentences entails the latter sentence.

The sentences are either formulae of the calculus or sentences of the four types mentioned above (p. 174 f.), of which the formulae of the calculus are molecular complexes.

Entailment is usually of the elementary type, where the implication-sentence of the conjunction of the entailing sentences, as antecedent, and the entailed sentence, as consequent, expresses a tautology in the Logic of Propositions.

Sometimes, however, entailment is of the type belonging to the Quantified Logic of Properties, as *e.g.*, when we say that $\bar{E}\ H\&A$ entails $H \overset{\bullet}{\subset} \bar{A}$ (p. 183), or of the type belonging to the Logic of Numbers (arithmetic), as *e.g.*, when we say that $1-p \geqslant 0$ entails $p \leqslant 1$ (p. 181).

One more type of entailment occurs occasionally, *viz.*, entailment in virtue of the following Principle of Identity or of Substitutability of Identities:

In a probability-expression which occurs in a formula we can substitute for the name of the first or the second property another name of the same property and for the name (numeral) of the real number another name of the same number. The substitution need not be performed in every place in the formula where the name occurs. It is in virtue of this Principle of Identity that we say, *e.g.*, that $P(A, H, p)$ and $p = r : q$ entail $P(A, H, r : q)$ (p. 182).

ii. Sometimes the deductive step takes the form of an assertion that from a certain formula of the calculus 'we get' a new formula in virtue of two 'technical' principles of formalized proofs, *viz.*

α. The Rule of Substitution: In a probability-expression which occurs in a formula we can substitute for the name of the first or the second property, and for the numeral, any other atomic or molecular name of a property from the same universe, and any other atomic or molecular name of a real number. The substitution must be performed in every place in the formula where the name occurs.

β. The Rule of Quantification: If we have the formula $\ldots \rightarrow \ldots$ and the antecedent but not the consequent contains the name x without being quantified in it, we also have the formula $(Ex)(\ldots) \rightarrow \ldots$. Conversely, if we have the formula $\ldots \rightarrow \ldots$ and the consequent but not the antecedent contains x without being quantified in it, we also have the formula $\ldots \rightarrow (x)(\ldots)$.

The Rule of Substitution is constantly used in the proofs. We shall not, however, make explicit reference to it. (The only exception is in the first proof.) The Rule of Quantification is not often used, and explicit reference is nearly always made to it in the context.

   iii.   In a few cases the deductive step is in virtue of the following Rule of Elimination:

If a probability-expression which occurs in a formula of the calculus is quantified in the numeral, then it can be omitted from (the perfect disjunctive normal form of) the formula.

The Rule of Elimination is always used in combination with the Rule of Quantification for the purpose of getting rid of 'auxiliary data.' (Cf. p. 180 and examples in §§6, 9, and 11.)[1]

Whereas the Principle of Identity and the Rules of Substitution and of Quantification are universal principles of formalized proofs, the Rule of Elimination seems to be peculiar to the Calculus of Probability. The need of the rule was pointed out by Reichenbach, who gave it a slightly narrower formulation.[2] I have not been able to do without it, nor to give it a more universal foundation in the logic of proof.[3]

The Logic of Propositions, the Logic of Properties and Arithmetic may conveniently be described as the logical sub-structure of our Calculus of Probability. The question of building up this sub-structure as an axiomatic and formalized deductive system does not concern us here.

*   *   *   *   *

The theorems or formulae which we want to prove have the same general structure as the axioms, *viz.*, they are implication-sentences, the consequents of which are themselves implication-sentences. The antecedent of a formula we shall call the first antecedent and the antecedent of the consequent the second

---

[1] That a probability-expression is quantified in the numeral presupposes that we have inserted, in virtue of the Rule of Quantification, either an existential operator in the antecedent or a universal operator in the consequent of a formula. The insertion of an operator again presupposes that the numeral in question occurs either only in the antecedent or only in the consequent. Thus, loosely speaking, the Rule of Elimination says that a probability-expression may be omitted from a formula, if the probability-value which its numeral names is not relevant to the numerical determination of other probability-values of the context.

[2] Reichenbach, *Wahrscheinlichkeitslehre*, §11. Reichenbach calls his rule the Rule of Existence.

[3] In my publication *Ueber Wahrscheinlichkeit* I used four 'extra-logical' rules of inference. Of these, however, the first three are included in the Rule of Elimination as stated here. The fourth can be formulated as an axiom, *viz.* the Axiom of Continuity.

antecedent. The consequent of a formula we shall call the first consequent and the consequent of the consequent we shall call the second consequent.

A convenient mode of conducting the proofs is this:

The sentences of which the first and second antecedents of the theorem we are going to prove, are molecular complexes, we call the 'data' of the problem. From the axioms and these data we want to deduce the second consequent of the theorem. The deduction is in steps of the three types mentioned above. It is easy to see that the deduction of the second consequent from the axioms and the data is equivalent to the deduction of the theorem from the axioms. (We need not show this in detail here.)

Sometimes 'auxiliary data' are needed. These always consist of a probability-expression. A theorem is proved in which the auxiliary datum occurs in the (second) antecedent. A combined use of the Rule of Quantification and the Rule of Elimination allows us to drop the auxiliary datum from the proved formula. Thus we finally get a theorem in which the auxiliary datum does not occur at all. (Cf. examples in §§6, 9, and 11.)

Sometimes also 'subsidiary data' are needed. These are always formal truths embodied in the sub-structure of the calculus, i.e., formal truths of the Logic of Propositions, the Logic of Properties, or the Logic of Numbers (arithmetic). These additional data do not occur explicitly in the proved theorems.

The following point should be observed: The 'higher' we ascend in the mathematical regions of probability, the more important do the above-mentioned subsidiary data of a mathematical nature become to the deduction of theorems. Some of the greatest achievements in the history of the subject, as e.g., the proofs of the various Principles of Great Numbers, have consisted, not so much in the computation of a new probability on the basis of given data, as in showing that a certain probability, thus computed, possesses such and such arithmetical properties, e.g., of attaining a maximum or converging towards a limit under certain conditions. In this inquiry, however, we are not interested in the arithmetical aspect, but only in the deductive connexions which trace the theorems back to the axioms of probability. The 'invasion' of the calculus by its mathematical sub-structure, which has been of primary importance for the development of probability mathematics, has sometimes tended to obscure these connexions in a way which has been fatal to the understanding of the philosophic implications of probability. This is particularly true of the notoriously obscure topic known as Inverse Probability, the clarification of which is one of the chief aims of the present treatment.

5. *The Maximum Principle*

According to $A2$, no probability, on existing evidence, is smaller than o. This is the Minimum Principle (Axiom). With the aid of $A4$ it is easy to prove from $A2$ that no probability on existing evidence is greater than 1. This may be called the Maximum Principle (Theorem).

The data are $E H$ and $P(A, H, p)$. We want to deduce $p \leqslant 1$.

In $A2$ we substitute $\overline{A}$ for A and $1-p$ for p. We get $(E H) \rightarrow (P(\overline{A}, H, 1-p) \rightarrow 1-p \geqslant 0)$.

The Addition Axiom $A4$ and the last formula entail, in the Logic of Propositions, $(E H) \rightarrow (P(A, H, p) \rightarrow 1-p \geqslant 0)$.

The data and the last formula entail, in the Logic of Propositions, $1-p \geqslant 0$.

$1-p \geqslant 0$ entails, in the Logic of Numbers (arithmetic), $p \leqslant 1$.

Thus we have proved

$T1 \ (E H) \rightarrow (P(A, H, p) \rightarrow p \leqslant 1)$.

This is the Maximum Theorem.

6. *The Multiplication Principle*

Given the additional data $p > 0$ and $q > 0$ respectively, the Multiplication Axiom $A5$ can be 'converted'. This means the following thing:

Given the data $E H\&A$ and $P(A, H, p)$ and $P(A\&B, H, r)$ and $p > 0$ we can deduce $P(B, H\&A, r : p)$, and given the data $E H\&A$ and $P(B, H\&A, q)$ and $P(A\&B, H, r)$ and $q > 0$ we can deduce $P(A, H, r : q)$.

As auxiliary datum for the first deduction we use

$P(B, H\&A, q)$ and as auxiliary datum for the second deduction we use $P(A, H, p)$.

A_5 and $E\ H\&A$ and $P(A, H, p)$ and $P(B, H\&A, q)$ entail, in the Logic of Propositions, $P(A\&B, H, pq)$.

$E\ H\&A$ entails $E\ H$.

A_1 and $E\ H$ and $P(A\&B, H, r)$ and $P(A\&B, H, pq)$ entail $r=pq$. $p>0$ and $r=pq$ entail, in the Logic of Numbers, $q=r:p$, and $q>0$ and $r=pq$ entail, in the Logic of Numbers, $p=r:q$.

In virtue of the Principle of Identity, $P(A, H, p)$ and $p=r:q$ entail $P(A, H, r:q)$, and $P(B, H\&A, q)$ and $q=r:p$ entail $P(B, H\&A, r:p)$.

Thus we have proved the formulae
$(E\ H\&A)\rightarrow(P(A, H, p)\&P(B, H\&A, q)\&P(A\&B, H, r)\&p>0\rightarrow P(B, H\&A, r:p))$ and $(E\ H\&A)\rightarrow(P(A, H, p)\&P(B, H\&A, q)\&P(A\&B, H, r)\&q>0\rightarrow P(A, H, r:q))$ respectively.

In virtue of the Rule of Quantification we get the new formulae
$(E\ H\&A)\rightarrow(P(A, H, p)\&(Eq)P(B, H\&A, q)\&P(A\&B, H, r)\&p>0\rightarrow P(B, H\&A, r:p))$ and $(E\ H\&A)\rightarrow((Ep)P(A, H, p)\&P(B, H\&A, q)\&P(A\&B, H, r)\&q>0\rightarrow P(A, H, r:q))$ respectively.[1]

In virtue of the Rule of Elimination, we can now drop the auxiliary data. So we finally get

T_2 $(E\ H\&A)\rightarrow(P(A, H, p)\&P(A\&B, H, r)\&p>0\rightarrow$
  $P(B, H\&A, r:p))$

and

T_3 $(E\ H\&A)\rightarrow(P(B, H\&A, q)\&P(A\&B, H, r)\&q>0\rightarrow$
  $P(A, H, r:q))$.

By the Multiplication Principle we shall understand the joint content of A_5 and T_2 and T_3.

* * * * *

Given the data $E\ H$ and $P(A, H, 0)$ we can deduce $P(A\&B, H, 0)$.

Two cases should be distinguished: either $H\&A$ exists or $H\&A$ is empty.

A_5 and $E\ H\&A$ and $P(A, H, 0)$ and $P(B, H\&A, q)$ entail, in the Logic of Propositions, $P(A\&B, H, 0q)$.

---

[1] It should be noted that $a\rightarrow(b\rightarrow c)$ is identical with $a\&b\rightarrow c$.

oq is identical with o. (This is an example of what we called above on p. 180 a 'subsidiary datum.')

In virtue of the Principle of Identity, $P(A\&B, H, oq)$ and $oq = o$ entail $P(A\&B, H, o)$.

Thus we have proved the formula

$(E\ H\&A)\rightarrow(P(A, H, o)\&P(B, H\&A, q)\rightarrow P(A\&B, H, o))$.

In virtue of the Rule of Quantification, we get from this

$(E\ H\&A)\rightarrow(P(A, H, o)\&(Eq)P(B, H\&A, q)\rightarrow P(A\&B, H, o))$.

In virtue of the Rule of Elimination, we get from this

$(E\ H\&A)\rightarrow(P(A, H, o)\rightarrow P(A\&B, H, o))$.

$\bar{E}\ H\&A$ entails, in the Logic of Properties, $H\subset\bar{A}$.

$A3$ and $E\ H$ and $H\subset\bar{A}$ entail, in the Logic of Propositions, $P(\bar{A}, H, 1)$.

$A4$ and $E\ H$ and $P(\bar{A}, H, 1)$ entail, in the Logic of Propositions, $P(A, H, o)$.

Further $\bar{E}\ H\&A$ also entails, in the Logic of Properties, $H\subset\overline{A\&B}$.

$A3$ and $E\ H$ and $H\subset\overline{A\&B}$ entail, in the Logic of Propositions, $P(\overline{A\&B}, H, 1)$.

$A4$ and $E\ H$ and $P(\overline{A\&B}, H, 1)$ entail, in the Logic of Propositions, $P(A\&B, H, o)$.

Thus we have proved the formula

$(E\ H)\&(\bar{E}\ H\&A)\rightarrow(P(A, H, o)\rightarrow P(A\&B, H, o))$.

The two formulae $(E\ H\&A)\rightarrow(P(A, H, o)\rightarrow P(A\&B, H, o))$ and $(E\ H)\&(\bar{E}\ H\&A)\rightarrow(P(A, H, o)\rightarrow P(A\&B, H, o))$ entail, in the Logic of Propositions,

$T4\ (E\ H)\rightarrow(P(A, H, o)\rightarrow P(A\&B, H, o))$.

Given the data $E\ H\&A$ and $P(B, H\&A, o)$ we can deduce $P(A\&B, H, o)$.

$A5$ and $E\ H\&A$ and $P(B, H\&A, o)$ and $P(A, H, p)$ entail, in the Logic of Propositions, $P(A\&B, H, po)$.

po is identical with o. (Subsidiary datum.)

In virtue of the Principle of Identity, $P(A\&B, H, po)$ and $po = o$ entail $P(A\&B, H, o)$.

Thus we have proved the formula

$(E\ H\&A)\rightarrow(P(A, H, p)\&P(B, H\&A, o)\rightarrow P(A\&B, H, o))$.

In virtue of the Rule of Quantification, we get from this
$(E\ H\&A)\rightarrow((Ep)P(A,\ H,\ p)\&P(B,\ H\&A,\ 0)\rightarrow P(A\&B,\ H,\ 0))$.

In virtue of the Rule of Elimination, we get from this
$T_5\ (E\ H\&A)\rightarrow(P(B,\ H\&A,\ 0)\rightarrow P(A\&B,\ H,\ 0))$.

T_4 and T_5 state that the conjunctive probability in A_5 is 0,
if either of the simple probabilities in A_5 is 0. It should be
observed that this value of the conjunctive probability is
independent not only of the *value* of the second of the simple
probabilities, but also of its *existence*.

7. The Principle of Extensionality

With the aid of the general principle of logic which we have
called the Principle of Identity it is possible to prove that the
probability-relation is extensional, meaning that not only
identical but also co-extensive properties are substitutable for
each other in the relation. We shall call this important principle
of probability the Principle of Extensionality. It is the joint
content of two theorems.

First we shall show that given the data $E\ H$ and $A\equiv B$ and
$P(A,\ H,\ p)$ we can deduce $P(B,\ H,\ p)$ and that given the data
$E\ H$ and $A\equiv B$ and $P(B,\ H,\ p)$ we can deduce $P(A,\ H,\ p)$.

Two cases should be distinguished: either $H\&A$ exists or
$H\&A$ is empty.

$E\ H\&A$ and $A\equiv B$ entail $E\ H\&B$. Thus $H\&B$ also exists.

$A\equiv B$ entails $H\&A\subset B$ and $H\&B\subset A$.

A_3 and $E\ H\&A$ and $H\&A\subset B$ entail $P(B,\ H\&A,\ 1)$.

A_3 and $E\ H\&B$ and $H\&B\subset A$ entail $P(A,\ H\&B,\ 1)$.

A_5 and $E\ H\&A$ and $P(A,\ H,\ p)$ and $P(B,\ H\&A,\ 1)$ entail
$P(A\&B,\ H,\ p)$.

$A\&B$ is identical with $B\&A$. (Subsidiary datum.)

In virtue of the Principle of Identity, $P(A\&B,\ H,\ p)$ and
$A\&B=B\&A$ entail $P(B\&A,\ H,\ p)$.

T_3 and $E\ H\&B$ and $P(A,\ H\&B,\ 1)$ and $P(B\&A,\ H,\ p)$ entail
$P(B,\ H,\ p)$.

Thus we have proved the formula
$(E\ H\&A)\&(A\equiv B)\rightarrow(P(A,\ H,\ p)\rightarrow P(B,\ H,\ p))$.

In an altogether similar manner we prove the formula
$(E\ H\&A)\&(A\equiv B)\rightarrow(P(B,\ H,\ p)\rightarrow P(A,\ H,\ p))$.

The two formulae entail the further formula

$(E\ H\&A)\&(A\equiv B)\rightarrow(P(A,\ H,\ p)\leftrightarrow P(B,\ H,\ p))$.

We have now to deal with the case when $H\&A$ is empty.

$\bar{E}\ H\&A$ and $A\equiv B$ entail $\bar{E}\ H\&B$. Thus $H\&B$ is also empty.

$\bar{E}\ H\&A$ entails $H\subset\bar{A}$ and $\bar{E}\ H\&B$ entails $H\subset\bar{B}$.

$A3$ and $A4$ and $E\ H$ and $H\subset\bar{A}$ entail $P(A,\ H,\ 0)$.

$A3$ and $A4$ and $E\ H$ and $H\subset\bar{B}$ entail $P(B,\ H,\ 0)$.

$A1$ and $E\ H$ and $P(A,\ H,\ p)$ and $P(A,\ H,\ 0)$ entail $p=0$.

In virtue of the Principle of Identity, $P(B,\ H,\ 0)$ and $p=0$ entail $P(B,\ H,\ p)$.

Thus we have proved the formula

$(E\ H)\&(\bar{E}\ H\&A)\rightarrow(P(A,\ H,\ p)\rightarrow P(B,\ H,\ p))$.

In an altogether similar manner we prove the formula

$(E\ H)\&(\bar{E}\ H\&A)\rightarrow(P(B,\ H,\ p)\rightarrow P(A,\ H,\ p))$.

The two formulae entail the further formula

$(E\ H)\&(\bar{E}\ H\&A)\rightarrow(P(A,\ H,\ p)\leftrightarrow P(B,\ H,\ p))$.

The formulae for the case when $H\&A$ exists and for the case when $H\&A$ is empty entail the general formula

$T6\ \ (E\ H)\&(A\equiv B)\rightarrow(P(A,\ H,\ p)\leftrightarrow P(B,\ H,\ p))$.

Secondly, we shall show that given the data $E\ H$ and $H\equiv G$ and $P(A,\ H,\ p)$ we can deduce $P(A,\ G,\ p)$ and that given the data $E\ H$ and $H\equiv G$ and $P(A,\ G,\ p)$ we can deduce $P(A,\ H,\ p)$.

The same two cases should be distinguished as before: either $H\&A$ exists or $H\&A$ is empty.

$E\ H\&A$ and $H\equiv G$ entail $E\ G\&A$. Thus $G\&A$ also exists.

$H\equiv G$ entails $H\&A\subset G$ and $G\&A\subset H$.

$A3$ and $E\ H\&A$ and $H\&A\subset G$ entail $P(G,\ H\&A,\ 1)$.

$A3$ and $E\ G\&A$ and $G\&A\subset H$ entail $P(H,\ G\&A,\ 1)$.

$E\ H$ and $H\equiv G$ entail $E\ G$.

$H\equiv G$ entails $H\subset G$ and $G\subset H$.

$A3$ and $E\ H$ and $H\subset G$ entail $P(G,\ H,\ 1)$.

$A3$ and $E\ G$ and $G\subset H$ entail $P(H,\ G,\ 1)$.

$A5$ and $E\ H\&A$ and $P(A,\ H,\ p)$ and $P(G,\ H\&A,\ 1)$ entail $P(A\&G,\ H,\ p)$.

$A\&G$ is identical with $G\&A$. (Subsidiary datum.)

In virtue of the Principle of Identity, $P(A\&G,\ H,\ p)$ and $A\&G=G\&A$ entail $P(G\&A,\ H,\ p)$.

$E\ H$ and $H\equiv G$ entail $E\ H\&G$.

T2 and *E H&G* and *P(G, H,* 1) and *P(G&A, H, p)* entail *P(A, H&G, p)*.

H&G is identical with *G&H*. (Subsidiary datum.)

In virtue of the Principle of Extensionality, *P(A, H&G, p)* and *H&G = G&H* entail *P(A, G&H, p)*.

A5 and *E H&G* and *P(H, G,* 1) and *P(A, G&H, p)* entail *P(H&A, G, p)*.

H&A is identical with *A&H*. (Subsidiary datum.)

In virtue of the Principle of Identity, *P(H&A, G, p)* and *H&A = A&H* entail *P(A&H, G, p)*.

T3 and *E G&A* and *P(H, G&A,* 1) and *P(A&H, G, p)* entail *P(A, G, p)*.

Thus we have proved the formula

$(E\ H\&A)\&(H\equiv G)\rightarrow(P(A, H, p)\rightarrow P(A, G, p))$.

In an altogether similar manner we prove the formula

$(E\ H\&A)\&(H\equiv G)\rightarrow(P(A, G, p)\rightarrow P(A, H, p))$.

The two formulae entail the further formula

$(E\ H\&A)\&(H\equiv G)\rightarrow(P(A, H, p)\longleftrightarrow P(A, G, p))$.

We have now to deal with the case when *H&A* is empty.

$\overline{E}\ H\&A$ and $H\equiv G$ entail $\overline{E}\ G\&A$. Thus *G&A* is also empty.

$\overline{E}\ H\&A$ entails $H\subset\overline{A}$ and $\overline{E}\ G\&A$ entails $G\subset\overline{A}$.

A3 and *A4* and *E H* and $H\subset\overline{A}$ entail *P(A, H,* 0).

A3 and *A4* and *E G* and $G\subset\overline{A}$ entail *P(A, G,* 0).

A1 and *E H* and *P(A, H, p)* and *P(A, H,* 0) entail *p = 0*.

In virtue of the Principle of Identity, *P(A, G,* 0) and *p = 0* entail *P(A, G, p)*.

Thus we have proved the formula

$(E\ H)\&(\overline{E}\ H\&A)\&(H\equiv G)\rightarrow(P(A, H, p)\rightarrow P(A, G, p))$.

In an altogether similar manner we prove the formula

$(E\ H)\&(\overline{E}\ H\&A)\&(H\equiv G)\rightarrow(P(A, G, p)\rightarrow P(A, H, p))$.

The two formulae entail the further formula

$(E\ H)\&(\overline{E}\ H\&A)\&(H\equiv G)\rightarrow(P(A, H, p)\longleftrightarrow P(A, G, p))$.

The formula for the case when *H&A* exists and the formula for the case when *H&A* is empty entail the general formula

T7 $(E\ H)\&(H\equiv G)\rightarrow(P(A, H, p)\longleftrightarrow P(A, G, p))$.

T6 and *T7* or the Principle of Extensionality having thus been proved, we shall no longer make explicit reference in the proofs to the Principle of Identity. This means that the inter-substitutability of identical, as opposed to 'merely' co-extensive, properties in the relation P is hereafter taken for granted.

Note.—In my publication *Ueber Wahrscheinlichkeit* (1945) I used the Principle of Identity only for numbers. In order to prove the Principle of Extensionality for the probability-relation I had to add to the system two more axioms, *viz.* $(E\,H)\rightarrow$ $(P(A\&B,\,H,\,p)\rightarrow P(B\&A,\,H,\,p))$ and $(E\,H\&G)\rightarrow(P(A,\,H\&G,\,p)\rightarrow P(A,\,G\&H,\,p))$. In no case need *T6* and *T7* themselves be taken as axioms.

8. *The Principle of Equivalence*

From the data $E\,H$ and $P(A,\,H,\,p)$ we can deduce $P(A\&H,\,H,\,p)$, and from the data $E\,H$ and $P(A\&H,\,H,\,p)$ we can deduce $P(A,\,H,\,p)$.

$H \subset H$ is tautologous. (Subsidiary datum.)

A3 and $E\,H$ and $H \subset H$ entail $P(H,\,H,\,1)$.

$E\,H$ entails $E\,H\&H$.

A5 and $E\,H\&H$ and $P(H,\,H,\,1)$ and $P(A,\,H\&H,\,p)$ entail $P(H\&A,\,H,\,p)$.

T2 and $E\,H\&H$ and $P(H,\,H,\,1)$ and $P(H\&A,\,H,\,p)$ entail $P(A,\,H\&H,\,p)$.

Thus we have proved the formula

T8 $(E\,H)\rightarrow(P(A,\,H,\,p)\leftrightarrow P(A\&H,\,H,\,p))$.

$A\&H \subset H$ is tautologous. *T8* thus means that any probability-expression can be replaced by another probability-expression in which the conjectured property is included in the field of measurement. The relation of probability between the conjectured and the evidential property is thus, in a certain sense, a relation between part and whole.

9. *The Addition Principle*

By the Addition Principle we shall understand the joint content of two theorems, called the Special and the General Addition Theorems respectively, together with their converses.

We shall first prove the Special Addition Theorem.

The data are $E\,H$ and $\overline{E}\,A\&B$ and $P(A,\,H,\,p)$ and $P(B,\,H,\,q)$. We want to deduce $P(A\text{v}B,\,H,\,p+q)$.

Two cases should be distinguished: either $H\&(A\text{v}B)$ exists or $H\&(A\text{v}B)$ is empty.

We shall first deal with the case when $H\&(AvB)$ exists.

As auxiliary datum we use $P(AvB, H, r)$.

Two sub-cases should be distinguished: either r is greater than o or r is o.

$(AvB)\&A$ is identical with A and $(AvB)\&B$ is identical with B.

$T2$ and $E\ H\&(AvB)$ and $P((AvB)\&A, H, p)$ and $P(AvB, H, r)$ and $r{>}o$ entail $P(A, H\&(AvB), p:r)$.

$T2$ and $E\ H\&(AvB)$ and $P((AvB)\&B, H, q)$ and $P(AvB, H, r)$ and $r{>}o$ entail $P(B, H\&(AvB), q:r)$.

$T8$ and $E\ H\&(AvB)$ and $P(B, H\&(AvB), q:r)$ entail $P(B\&H\&(AvB), H\&(AvB), q:r)$.

$\overline{E}\ A\&B$ entails $B\&H\&(AvB)\equiv\overline{A}\&H\&(AvB)$.

$T6$ and $E\ H\&(AvB)$ and $B\&H\&(AvB)\equiv\overline{A}\&H\&(AvB)$ and $P(B\&H\&(AvB), H\&(AvB), q:r)$ entail $P(\overline{A}\&H\&(AvB), H\&(AvB), q:r)$.

$T8$ and $E\ H\&(AvB)$ and $P(\overline{A}\&H\&(AvB), H\&(AvB), q:r)$ entail $P(\overline{A}, H\&(AvB), q:r)$.

$A4$ and $E\ H\&(AvB)$ and $P(A, H\&(AvB), p:r)$ entail $P(\overline{A}, H\&(AvB), 1{-}p:r)$.

$A1$ and $E\ H\&(AvB)$ and $P(\overline{A}, H\&(AvB), q:r)$ and $P(\overline{A}, H\&(AvB), 1{-}p:r)$ entail $q:r=1{-}p:r$.

$r{>}o$ and $q:r=1{-}p:r$ entail $r=p+q$.

$P(AvB, H, r)$ and $r=p+q$ entail $P(AvB, H, p+q)$.

We shall now deal with the sub-case $r=o$.

$T4$ and $E\ H$ and $P(AvB, H, o)$ entail $P((AvB)\&A, H, o)$ and $P((AvB)\&B, H, o)$.

Thus we have $P(A, H, o)$ and $P(B, H, o)$.

$A1$ and $E\ H$ and $P(A, H, p)$ and $P(A, H, o)$ entail $p=o$.

$A1$ and $E\ H$ and $P(B, H, q)$ and $P(B, H, o)$ entail $q=o$.

$r=o$ and $p=o$ and $q=o$ entail $r=p+q$.

Thus we have proved the formula

$(E\ H)\&(\overline{E}\ A\&B)\&(E\ H\&(AvB))\rightarrow$
$(P(A, H, p)\&P(B, H, q)\&P(AvB, H, r)\rightarrow P(AvB, H, p+q))$.

In virtue of the Rule of Quantification, we can replace $P(AvB, H, r)$ in the formula by $(Er)P(AvB, H, r)$.

In virtue of the Rule of Elimination, $(Er)P(AvB, H, r)$ can be dropped from the formula. Thus we get the formula

$(E\ H)\&(\bar{E}\ A\&B)\&(E\ H\&(AvB))\to(P(A,\ H,\ p)\&P(B,\ H,\ q)\to P(AvB,\ H,\ p+q))$.

We shall now deal with the case when $H\&(AvB)$ is empty. No auxiliary datum is needed.

$\bar{E}\ H\&(AvB)$ entails $H\subset\overline{AvB}$ and $H\subset\bar{A}$ and $H\subset\bar{B}$.

A_3 and A_4 and $E\ H$ and $H\subset\overline{AvB}$ and $H\subset\bar{A}$ and $H\subset\bar{B}$ entail $P(AvB,\ H,\ 0)$ and $P(A,\ H,\ 0)$ and $P(B,\ H,\ 0)$.

A_1 and $E\ H$ and $P(A,\ H,\ 0)$ and $P(A,\ H,\ p)$ entail $p=0$.

A_1 and $E\ H$ and $P(B,\ H,\ 0)$ and $P(B,\ H,\ q)$ entail $q=0$.

$p=0$ and $q=0$ entail $p+q=0$.

$P(AvB,\ H,\ 0)$ and $p+q=0$ entail $P(AvB,\ H,\ p+q)$.

Thus we have proved the formula

$(E\ H)\&(\bar{E}\ A\&B)\&(\bar{E}\ H\&(AvB))\to(P(A,\ H,\ p)\&P(B,\ H,\ q)\to P(AvB,\ H,\ p+q))$.

The formula for the case when $H\&(AvB)$ exists and the formula for the case when $H\&(AvB)$ is empty entail the general formula

T_9 $(E\ H)\&(\bar{E}\ A\&B)\to(P(A,\ H,\ p)\&P(B,\ H,\ q)\to$
$\quad P(AvB,\ H,\ p+q))$.

This is the Special Addition Theorem. There are two 'converses' of it which we give here without proof, *viz.*

T_{10} $(E\ H)\&(\bar{E}\ A\&B)\to(P(A,\ H,\ p)\&P(AvB,\ H,\ q)\to$
$\quad P(B,\ H,\ q\text{-}p))$

and

T_{11} $(E\ H)\&(\bar{E}\ A\&B)\to(P(B,\ H,\ p)\&P(AvB,\ H,\ q)\to$
$\quad P(A,\ H,\ q\text{-}p))$.

We shall next prove the General Addition Theorem.

The data are $E\ H$ and $P(A,H,p)$ and $P(B,H,q)$ and $P(A\&B,H,r)$. We want to deduce $P(AvB,H,p+q-r)$.

Two cases should be distinguished: either $H\&B$ exists or $H\&B$ is empty.

We shall first deal with the case when $H\&B$ exists.

Two sub-cases should be distinguished: either q is greater than 0 or q is 0.

T_2 and $E\ H\&B$ and $P(B,\ H,\ q)$ and $P(A\&B,\ H,r)$ and $q>0$ entails $P(A,\ H\&B,\ r:q)$.

A_4 and $E\ H\&B$ and $P(A,\ H\&B,\ r:q)$ entail
$P(\bar{A},\ H\&B,\ 1\text{-}r:q)$.

A_5 and $E\ H\&B$ and $P(B, H, q)$ and $P(\bar{A}, H\&B, 1-r:q)$ and $q>0$ entail $P(B\&\bar{A}, H, q\text{-}r)$.

$\bar{E}\ A\&B\&\bar{A}$ is tautologous.

T_9 and $E\ H$ and $\bar{E}\ A\&B\&\bar{A}$ and $P(A, H, p)$ and $P(B\&\bar{A}, H, q\text{-}r)$ entail $P(A\lor B\&\bar{A}, H, p+q-r)$.

$A\lor B\&\bar{A}$ is identical with $A\lor B$.

Thus we have $P(A\lor B, H, p+q—r)$.

We shall now deal with the sub-case $q=0$.

T_4 and $E\ H$ and $P(B, H, 0)$ entail $P(A\&B, H, 0)$ and $P(B\&\bar{A}, H, 0)$.

A_1 and $E\ H$ and $P(A\&B, H, r)$ and $P(A\&B, H, 0)$ entail $r=0$.

$q=0$ and $r=0$ entail $q\text{-}r=0$.

$P(B\&\bar{A}, H, 0)$ and $q\text{-}r=0$ entail $P(B\&\bar{A}, H, q\text{-}r)$.

As we already know, T_9 and $E\ H$ and $\bar{E}\ A\&B\&\bar{A}$ and $P(A, H, p)$ and $P(B\&\bar{A}, H, q\text{-}r)$ entail $P(A\lor B, H, p+q-r)$.

It remains for us to deal with the case when $H\&B$ is empty.

$\bar{E}\ H\&B$ entails $H\subset\bar{B}$.

A_3 and A_4 and $E\ H$ and $H\subset\bar{B}$ entail $P(B, H, 0)$.

$P(A\lor B, H, p+q-r)$ can now be deduced in exactly the same way as under the alternative $q=0$.

We have herewith completed the proof of the formula

T_{12} $(E\ H)\to(P(A, H, p)\&P(B, H, q)\&P(A\&B, H, r)$
$\to P(A\lor B, H, p+q-r))$

This is the General Addition Theorem. There are three 'converses' of it which we give here without proof, *viz.*

T_{13} $(E\ H)\to(P(B, H, p)\&P(A\&B, H, q)\&P(A\lor B, H, r)\to$
$P(A, H, q+r-p))$

T_{14} $(E\ H)\to(P(A, H, p)\&P(A\&B, H, q)\&P(A\lor B, H, r)\to$
$P(B, H, q+r-p))$

and

T_{15} $(E\ H)\to(P(A, H, p)\&P(B, H, q)\&P(A\lor B, H, r)\to$
$P(A\&B, H, p+q-r))$.

10. *The Inverse Principle*

There are three data as to the existence of fields of measurement, *viz.* $E\ H\&A$ and $E\ H\&B$ and $E\ H\&C$. The properties

A and *B* are mutually exclusive, *i.e.*, $\overline{E}$ *A*&*B*. The same properties are further jointly exhaustive of the field *H*&*C*, *i.e.* *H*&*C* $\subset$ *A*v*B*. There are four data as to the existence of probabilities, *viz.* $P(A, H, p)$ and $P(B, H, q)$ and $P(C, H\&A, r)$ and $P(C, H\&B, s)$. Finally there is the datum $pr+qs>$0.

On these data we want to deduce $P(A, H\&C, pr : (pr+qs))$ and $P(B, H\&C, qs : (pr+qs))$.

A5 and *E* *H*&*A* and $P(A, H, p)$ and $P(C, H\&A, r)$ entail $P(A\&C, H, pr)$.

A5 and *E* *H*&*B* and $P(B, H, q)$ and $P(C, H\&B, s)$ entail $P(B\&C, H, qs)$.

$\overline{E}$ *A*&*B* entails $\overline{E}$ *A*&*B*&*C*.

E *H*&*A* entails *E* *H*.

T9 and *E* *H* and $\overline{E}$ *A*&*B*&*C* and $P(A\&C, H, pr)$ and $P(B\&C, H, qs)$ entail $P(A\&CvB\&C, H, pr+qs)$.

T8 and *E* *H* and $P(A\&CvB\&C, H, pr+qs)$ entail $P((A\&Cv B\&C)\&H, H, pr+qs)$.

H&*C* $\subset$ *A*v*B* entails $(A\&CvB\&C)\&H\equiv C\&H$.

T6 and *E* *H* and $(A\&CvB\&C)\&H\equiv C\&H$ and $P((A\&Cv B\&C)\&H, H, pr+qs)$ entail $P(C\&H, H, pr+qs)$.

T8 and *E* *H* and $P(C\&H, H, pr+qs)$ entail $P(C, H, pr+qs)$.

T2 and *E* *H*&*C* and $P(A\&C, H, pr)$ and $P(C, H, pr+qs)$ and $pr+qs>$0 entail $P(A, H\&C, pr : (pr+qs))$.

T2 and *E* *H*&*C* and $P(B\&C, H, qs)$ and $P(C, H, pr+qs)$ and $pr+qs>$0 entail $P(B, H\&C, qs : (pr+qs))$.

Thus we have proved the formula

T16 $(E$ *H*&*A*$)$&$(E$ *H*&*B*$)$&$(E$ *H*&*C*$)$&$(\overline{E}$ *A*&*B*$)$& $(H\&C \subset AvB)\rightarrow(P(A, H, p)\&P(B, H, q)\&P(C, H\&A, r)\& P(C, H\&B, s)\&pr+qs>0\rightarrow P(A, H\&C, pr : (pr+qs)))$

and the symmetrical formula

T17 $(E$ *H*&*A*$)$&$(E$ *H*&*B*$)$&$(E$ *H*&*C*$)$&$(\overline{E}$ *A*&*B*$)$& $(H\&C \subset AvB)\rightarrow(P(A, H, p)\&P(B, H, q)\&P(C, H\&A, r)\& P(C, H\&B, s)\&pr+qs>0P(B, H\&C, qs : (pr+qs)))$.

11. The Composition Principle

The data are $E\,H$ and $E\,G$ and $\bar{E}\,H\&G$ and $P(A, H, p)$ and $P(A, G, q)$ and $P(H, HvG, r)$. We want to deduce $P(A, HvG, pr+q-qr)$.

$E\,H$ entails $E\,HvG$.

$HvG \subset HvG$ is tautologous.

A_3 and $E\,HvG$ and $HvG \subset HvG$ entails $P(HvG, HvG, 1)$.

T_{10} and $E\,HvG$ and $\bar{E}\,H\&G$ and $P(HvG, HvG, 1)$ and $P(H, HvG, r)$ entail $P(G, HvG, 1\text{-}r)$.

H is identical with $H\&(HvG)$ and G is identical with $G\&(HvG)$.

A_5 and $E\,H$ and $P(H, HvG, r)$ and $P(A, H\&(HvG), p)$ entail $P(A\&H, HvG, pr)$.

A_5 and $E\,G$ and $P(G, HvG, 1\text{-}r)$ and $P(A, G\&(HvG), q)$ entail $P(A\&G, HvG, q\text{-}qr)$.

$\bar{E}\,H\&G$ entails $\bar{E}\,A\&H\&G$.

T_9 and $E\,HvG$ and $\bar{E}\,A\&H\&G$ and $P(A\&H, HvG, pr)$ and $P(A\&G, HvG, q\text{-}qr)$ entail $P(A\&HvA\&G, HvG, pr+q-qr)$.

$A\&HvA\&G$ is identical with $A\&(HvG)$.

T_8 and $E\,HvG$ and $P(A\&(HvG), HvG, pr+q-qr)$ entail $P(A, HvG, pr+q-qr)$.

Thus we have proved the formula

T_{18} $(E\,H)\&(E\,G)\&(\bar{E}\,H\&G)\rightarrow$
$\qquad(P(A, H, p)\&P(A, G, q)\&P(H, HvG, r)\rightarrow P(A, HvG, pr+q-qr))$.

This we shall call the General Composition Theorem. There are three 'converses' of it which we give without proof, *viz.*

T_{19} $(E\,H)\&(E\,G)\&(\bar{E}\,H\&G)\rightarrow$
$\qquad(P(A, G, p)\&P(H, HvG, q)\&P(A, HvG, r)\&q>0\rightarrow$
$\qquad P(A, H, (r-p+pq):q))$

T_{20} $(E\,H)\&(E\,G)\&(\bar{E}\,H\&G)\rightarrow$
$\qquad(P(A, H, p)\&P(H, HvG, q)\&P(A, HvG, r)\&q<1\rightarrow$
$\qquad P(A, G, (r\text{-}pq):(1\text{-}q)))$

T_{21} $(E\,H)\&(E\,G)\&(\bar{E}\,H\&G)\rightarrow$
$\qquad(P(A, H, p)\&P(A, G, q)\&P(A, HvG, r)\&p\text{-}q>0\rightarrow$
$\qquad P(H, HvG, (r\text{-}q):(p\text{-}q)))$.

The datum $p\text{-}q>0$ requires a comment. If p equals q, then the probability of H on evidence HvG cannot be computed from the data. It does not follow that it does not exist. If p is smaller than q, then, since the data are 'symmetrical' in H and G, we can compute the probability of G on evidence HvG. From this we get the probability of H on evidence HvG by subtracting the computed value from 1.

In the proof of $T18$, let the datum $P(A, G, q)$ be replaced by $P(A, G, p)$.

Then we obtain the formula $(E\,H)\&(E\,G)\&(\overline{E}\,H\&G)\to$
$(P(A, H, p)\&P(A, G, p)\&P(H, HvG, r)\to P(A, HvG, p))$.

In virtue of the Rule of Quantification, we get from this
$(E\,H)\&(E\,G)\&(\overline{E}\,H\&G)\to$
$(P(A, H, p)\&P(A, G, p)\&(Er)P(H, HvG, r)\to P(A, HvG, p))$.

In virtue of the Rule of Elimination, we finally get

$T22$ $(E\,H)\&(E\,G)\&(\overline{E}\,H\&G)\to(P(A, H, p)\&P(A, G, p)\to$
    $P(A, HvG, p))$.

This we shall call the Special Composition Theorem.

12. Independence

If $H\&B$ exists and if the probability of A on evidence $H\&B$ is the same as the probability of A on evidence H alone, then we call A independent (for probability) of B in H.

If A is independent of B in H, then we say that B is irrelevant (for probability) to A in H.

If A is independent of B in H and B is independent of A in H, then we call A and B mutually independent in H.

The data $E\,H\&B$ and $P(A, H, p)$ and $P(A, H\&B, p)$ thus make A independent of B in H.

On the additional data $E\,H\&A$ and $p>0$ and $P(B, H, q)$ we easily deduce $P(B, H\&A, q)$.

If we replace the additional datum $P(B, H, q)$ by $P(A\&B, H, q)$, we easily deduce $P(B, H, q:p)$ and $P(B, H\&A, q:p)$.

If we replace the additional datum $P(B, H, q)$ by $P(B, H\&A, q)$, we easily deduce $P(B, H, q)$.

Thus on the above three alternatives as to additional data we can, from the independence of A of B in H, conclude the mutual independence of the two properties in H.

It is of some importance to observe that the conclusion from independence to mutual independence requires additional data.

If A is independent of B and also of $\bar{B}$ in H, then we call A completely independent of B in H (or B completely irrelevant to A in H).

As we know, the data $E\ H\&B$ and $P(A, H, p)$ and $P(A, H\&B, p)$ constitute the independence of A of B in H.

The additional data $E\ H\&\bar{B}$ and $P(B, H, q)$ and $q<1$ make A completely independent of B in H. The proof is as follows:

$T20$ and $E\ H\&B$ and $E\ H\&\bar{B}$ and $\bar{E}\ H\&B\&\bar{B}$ and $P(A, H\&B, p)$ and $P(B, H\&Bv H\&\bar{B}, q)$ and $P(A, H\&Bv H\&\bar{B}, p)$ and $q<1$ entail $P(A, H\&\bar{B}, p)$. Thus A is completely independent of B in H.

It is of some importance to observe that the conclusion from independence to complete independence requires additional data.

If A and B are mutually independent in H, then the probability of $A\&B$ on evidence H is the probability of A on evidence H multiplied by the probability of B on evidence H. This is sometimes referred to as the Special Multiplication Theorem.

13. The Generalized Elementary Principles

We have now completed our contribution to the building up of the first stage of the Calculus of Probability. We proceed to the second stage.

Our first task will be to generalize the Multiplication, Addition, Inverse and (Special) Composition Principles of the first stage so that they may become valid for any products and sums of any number of members of sequences of properties.

The generalization is a trivial matter of reasoning from n to $n+1$.

In order to reach full generality we have to express the theorems in the rather complicated symbolism of index numbers which we introduced in Chap. II, §5. For some purposes, however, it will be sufficient to use simpler theorems which are valid, not for *any* products and sums of n members of a sequence of properties, but for products and sums of the *n first* members of a sequence of properties.

Formalized proofs of full generality are laborious to work out. Since they differ from the corresponding proofs of the first stage of the calculus merely in two 'trivial' features, *viz.*

i. in introducing subsidiary data (p. 180) concerning index numbers, and

ii. in containing an inductive step from n to $n+1$, we shall not burden our exposition with the proofs.[1]

We shall, for the sake of convenience, try to make as little use of the symbolism as possible.

A. The Multiplication Principle.

Let there be a property H and a sequence of properties $A_1, \ldots, A_n, \ldots$

For a random (positive) product of m members of the A-sequence we have the name ^nK_m. (Cf. above p. 58.)

The data are as follows:

The product of H and any product ^nK_m is not empty.

Any member A_i of the A-sequence has a probability p_i on evidence H.

Any member A_i of the A-sequence also has a probability $^np_i^m$ on the product of H *and* any product ^nK_m as evidence.

On these data we can compute a probability for any product ^nK_m on evidence H. This probability is the probability of the first A-property in ^nK_m on H alone as evidence, multiplied by the probability of the second A-property in ^nK_m on H *and* the first A-property in ^nK_m as evidence, multiplied by . . ., multiplied by the probability of the last A-property in ^nK_m on H *and* the product of all other A-properties in ^nK_m as evidence.

Given m and n, the indices of the A-properties in ^nK_m are uniquely determined. So also are the indices of any product of some A-properties in ^nK_m. Thus, given m and n, the indices of the probabilities, the product of which is the computed probability of ^nK_m on evidence H, are uniquely determined. The computed probability can therefore be called ^np_m. (By convention $^np_1$ should be interpreted as p_n.)

[1] There is a more formalized exposition in my publication *Ueber Wahrscheinlichkeit*.

Thus the generalized form of the Multiplication Axiom $A5$ can be expressed in symbols as follows:

$T23$ (m) (n) $(E\ H\&^nK_m) \rightarrow$
$\qquad ((i)P(A_i, H, p_i)\&(i)(m)(n)P(A_i, H\&^nK_m, {}^np_i^m) \rightarrow$
$\qquad (m)(n)P({}^nK_m, H, {}^np_m)).$

We shall not write down in symbols the two 'converses' corresponding to $T2$ and $T3$.

For many purposes it will be sufficient to know, not the probability of *any* product of A-properties, but the probability of the product of the n *first* A-properties. For these purposes we can use the simpler theorem

$T23'$ $(n)(E\ H\&\Pi A_n) \rightarrow (P(A_1, H, p_1)\&(n)P(A_{n+1}, H\&\Pi A_n, p_{n+1})$
$\qquad \rightarrow (n)P(\Pi A_n, H, \Pi p_n)).$

B. The Addition Principle.

Let there be a property H and a sequence of properties $A_1, \ldots, A_n, \ldots$.

For a random (positive) sum of m members of the A-sequence we have the name ${}^nM_m$. (Cf. above p. 58.)

The data are as follows:

The property H is not empty.

Any two members of the A-sequence are mutually exclusive.

Any member A_i of the A-sequence has a probability p_i on evidence H.

On these data we can compute a probability for any sum ${}^nM_m$ on evidence H. This probability is the sum ${}^n\sigma_m$ (cf. above p. 61) of the probabilities, on evidence H, of the m A-properties in the sum ${}^nM_m$.

Thus the generalized form of the Special Addition Theorem $T9$ can be expressed in symbols as follows:

$T24$ $(E\ H)\&(m)(n)(\overline{m=n} \rightarrow (\overline{E}\ A_m\&A_n)) \rightarrow ((i)P(A_i, H, p_i) \rightarrow$
$\qquad (m)(n)P({}^nM_m, H, \sigma^n_m)).$

We shall not write down in symbols the two 'converses' corresponding to $T10$ and $T11$.

For the sum of the n *first* A-properties the theorem runs:

$T24'$ $(E\ H)\&(m)(n)(\overline{m=n} \rightarrow (\overline{E}\ A_m\&A_n)) \rightarrow ((i)P(A_i, H, p_i) \rightarrow$
$\qquad (n)P(\Sigma A_n, H, \Sigma p_n)).$

Replace the datum that any two members of the A-sequence are mutually exclusive by the datum that any product ${}^nK_m$ has a probability ${}^np_m$ on evidence H. This new datum entails as a limiting case the above datum, that any member A_i of the A-sequence has a probability p_i on evidence H.

On these more general data we can also compute a probability for any sum ^nM_m on evidence H. This probability equals the sum $^n\sigma_m$ or $^n_1\theta^m$ (cf. above p. 61) of the probabilities on evidence H of the m A-properties in ^nM_m, minus the sum $^n_2\theta^m$ of the probabilities on evidence H of the products of any two A-properties in ^nM_m, plus the sum $^n_3\theta^m$ of the probabilities on evidence H of the products of any three A-properties in ^nM_m minus . . ., ^np_m or $^n_m\theta^m$, or the probability on evidence H of the product of all the A-properties in ^nM_m.

Thus the generalized form of the General Addition Theorem $T12$ can be expressed in symbols as follows:

$T25$ $(E\ H) \rightarrow ((m)(n)P(^nK_m,\ H,\ ^np_m) \rightarrow$
$\qquad (m)(n)P(^nM_m,\ H,\ ^n_1\theta^m\text{-}^n_2\theta^m+^n_3\theta^m\text{-}\ .\ .\ .\ ^n_m\theta^m)).$

We shall not write down in symbols the three 'converses' corresponding to $T13$, $T14$, and $T15$.

C. The Inverse Principle.

Let there be two properties B and H and a sequence of properties $A_1, \ldots, A_n, \ldots$.

The data are as follows:

The product of H and B is not empty.

The product of H and any member A_i of the A-sequence is not empty.

Any two members of the A-sequence are mutually exclusive.

Any member A_i of the A-sequence has a probability p_i on evidence H.

The property B has a probability q_i on the product of H *and* any member A_i of the A-sequence as evidence.

The product $H\&B$ is included in the sum ^nM_m.

A_i is one of the A-properties in the sum ^nM_m. This we express $Inc(n, m, i, 1)$. (Cf. above p. 60.)

On these data we can compute a probability for A_i on evidence $H\&B$. This probability equals the product of the probability of A_i on evidence H and the probability of B on evidence $H\&A_i$ divided by the sum $_n\Omega_m$ (cf. above p. 62) of all such products, when A_i is in order the first, the second, . . ., the last A-property in the sum ^nM_m. It is assumed that $_n\Omega_m$ is greater than 0.)

Thus the generalized form of the Inverse Theorem $T16$ (and $T17$) can be expressed in symbols as follows:

$T26$ $(E\ H\&B)\&(i)(E\ H\&A_i)\&(m)(n)(\overline{m=n}\rightarrow(\bar{E}\ A_m\&A_n))\rightarrow$
$\quad((i)P(A_i,H,p_i)\&(i)P(B,H\&A_i,q_i)\&H\&B\subset{}^nM_m\&Inc(n,m,i,1)$
$\quad\&{}_n\Omega_m>0\rightarrow P(A_i,\ H\&B,\ p_iq_i:{}_n\Omega_m)).$

Replace the datum above that A_i is one of the A-properties in the sum ${}^nM_m$ by the new datum that ${}^kM_g$ is a sum of g of the m A-properties in the sum ${}^nM_m$. This we express $Inc(n, m, k, g)$.

The probability, on evidence $H\&B$, of each one of the A-properties in the sum ${}^kM_g$ can be computed from $T26$. Since the A-properties are mutually exclusive, the probability on evidence $H\&B$ of the sum ${}^kM_g$ itself can be computed from $T24$. In this way we get a new theorem which we call the Extended Inverse Theorem. In symbols:

$T27$ $(E\ H\&B)\&(i)(E\ H\&A_i)\&(m)(n)(\overline{m=n}\rightarrow(\bar{E}\ A_m\&A_n))\rightarrow$
$\quad((i)P(A_i,H,p_i)\&(i)P(B,H\&A_i,q_i)\&H\&B\subset{}^nM_m\&Inc(n,m,k,g)$
$\quad\&{}_n\Omega_m>0\rightarrow P({}^kM_g,\ H\&B,\ {}_k\Omega_g:{}_n\Omega_m)).$

D. The Composition Principle.

A generalization of the General Composition Principle is not needed. ·

Let there be a property A and a sequence of properties $H_1,\ldots,H_n,\ldots$.

For a random (positive) sum of m members of the H-sequence we shall use the name ${}^nM_m$.

The data are as follows:

Every one of the H-properties exists.

Any two of the H-properties are mutually exclusive.

The property A has a probability p on any H_i as evidence.

On these data it can be shown that the property A has the same probability p also on any sum ${}^nM_m$ as evidence.

Thus the generalized form of the Special Composition Theorem $T22$ can be expressed in symbols as follows:

$T28$ $(i)(E\ H_i)\&(m)(n)(\overline{m=n}\rightarrow(\bar{E}\ H_m\&H_n))\rightarrow((i)P(A,\ H_i,\ p)\rightarrow$
$\quad(m)(n)P(A,\ {}^nM_m,\ p)).$

14. Independence-Realms

Our next task will be to generalize the concept of independence and to introduce the important notion of Independence-Realms.

The members of the sequence $A_1, \ldots, A_n, \ldots$, of properties are said to be totally independent for probability in H, if, and only if, the following three conditions are fulfilled:

    i.   The product of H *and* any (positive) product of members of the sequence is not empty. In symbols:

(1) $(m)(n)(E\ H\&{}^n K_m)$.

    ii. Every one of the members of the sequence has a probability on H as evidence and also on the product of H *and* any (positive) product of other members of the sequence as evidence. In symbols:

(2) $(i)P(A_i, H, p_i)\ \&\ (i)(m)(n)(Exc(n, m, i,\ 1){\rightarrow}P(A_i, H\&{}^n K_m, {}^n p_i^m))$.

    iii. The probability of any member of the sequence on H as evidence equals the probability of the same member on the product of H *and* any (positive) product of other members of the sequence as evidence. In symbols:

(3) $(i)(m)(n)(Exc(n, m, i,\ 1){\rightarrow}p_i={}^n p_i^m)$.

The sequence $A_1, \ldots, A_n, \ldots$ of properties is said to constitute an Independence-Realm in H, if, and only if, we retain conditions ii and iii above and replace i by the stronger condition:

    iv. The product of H *and* any product or sum of some $m(0 \leqslant m \leqslant n)$ of some $n(1 \leqslant n)$ members of the sequence and the negation-properties of the remaining $n{-}m$ members is not empty. In symbols (considering that the negation of a conjunction can be expressed as a disjunction in virtue of the Laws of de Morgan):

(4) $(g)(k)(m)(n)(Exc(g, m, k, n{-}m){\rightarrow}(E\ H\&{}^g K_m\&{}^k L_{n\text{-}m})\ \&$
    $(E\ H\&{}^g K_m\&{}^k L_{n\text{-}m}))$.

The sequence $A_1, \ldots, A_n, \ldots$ of properties is said to constitute a Normal Independence-Realm in H, if, and only if, we add to ii and iii and iv the following condition :

v. The probability of any member of the sequence on evidence H is not extreme. In symbols:

(5) $(i)(p_i > 0 \, \& \, p_i < 1)$

If the sequence $A_1, \ldots, A_n, \ldots$ of properties constitutes a Normal Independence-Realm in H, we can prove the following ten theorems:[1]

1. Any two positive products with no common constituent are mutually independent in H.

Proof: From ii and iv in combination with the Multiplication Principle it follows that ^nK_m has a probability ^np_m on evidence H. From the condition of independence iii it follows that ^np_m equals the product of the probabilities on evidence H of the m A-properties in ^nK_m. For this product we have the name $^n\pi_m$. (Cf. above p. 61.)

If ^nK_m and ^kK_g have no common constituent, then from ii and iv in combination with the Multiplication Principle it follows that ^nK_m has a probability ^nq_m on evidence $H \& {}^kK_g$. From the condition of independence iii it follows also that ^nq_m equals $^n\pi_m$.

Thus $^np_m = {}^nq_m$ and ^nK_m is independent of ^kK_g in H.

In a similar manner we prove that ^kK_g is independent of ^nK_m in H.

Thus ^nK_m and ^kK_g are mutually independent in H.

2. The negations of the properties $A_1, \ldots, A_n, \ldots$ are totally independent in H.

Proof: Since there is a probability ^np_m of ^nK_m on evidence H, it follows from the General Addition Principle that there is a probability $^n_1\theta^m - {}^n_2\theta^m + {}^n_3\theta^m - \ldots {}^n_m\theta^m$ of ^nM_m (cf. above p. 61) on evidence H. But ^np_m equals $^n\pi_m$. Consequently (cf. above p. 61) $^n_1\theta^m - {}^n_2\theta^m + {}^n_3\theta^m - \ldots {}^n_m\theta^m$ equals $^n_1\Pi^m - {}^n_2\Pi^m + {}^n_3\Pi^m - \ldots {}^n_m\Pi^m$. It follows from the Addition Axiom that the probability of $\overline{^nM_m}$ on evidence H is $1 - {}^n_1\Pi^m + {}^n_2\Pi^m - {}^n_3\Pi^m + \ldots {}^n_m\Pi^m$.

According to the Laws of de Morgan $\overline{^nM_m}$ is identical with ^nL_m (cf. above p. 58). By simple arithmetical considerations it

---

[1] A statement in symbols will not be given here.

can be shown that $1 - {}^n_1\Pi^m + {}^n_2\Pi^m - {}^n_3\Pi^m + \ldots {}^n_m\Pi^m$ is identical with ${}^n\rho_m$ (cf. above p. 61). Thus the probability of ${}^nL_m$ on evidence H is ${}^n\rho_m$. It follows from v that ${}^n\rho_m$ is greater than 0.

If A_i is not a constituent of ${}^nM_m$, it follows from 1 and the General Addition Principle and the Addition Axiom that the probability of ${}^nL_m$ on evidence $H\&A_i$ is also ${}^n\rho_m$.

It follows from the Multiplication Axiom that the probability of $A_i\&{}^nL_m$ on evidence H is $p_i \cdot {}^n\rho_m$. Since ${}^n\rho_m>0$, it follows from the first converse ($T2$) of the Multiplication Axiom that the probability of A_i on evidence $H\&{}^nL_m$ is p_i. It follows from the Addition Axiom that the probability of $\overline{A_i}$ on evidence $H\&{}^nL_m$ is $1-p_i$.

It follows from ii that the probability of $\overline{A_i}$ on evidence H alone is also $1-p_i$.

Thus according to our definition of total independence, the negations of the A-properties are totally independent in H.

3. Any two negative products with no common constituent are mutually independent in H.

After the proof of 2 we can prove 3 in a way which is closely similar to the proof of 1.

4. Any negative product is independent of any positive product in H supposing the two products do not contradict one another.

Proof: As we have seen, the probability of ${}^nL_m$ on evidence H is ${}^n\rho_m$. But ${}^nL_m$ is identical with $\overline{{}^nM_m}$. Consequently, the probability of ${}^nM_m$ on evidence H is $1-{}^n\rho_m$. If ${}^nL_m$ and ${}^kK_g$ do not contradict one another, it follows from 1 and the Special Addition Theorem that the probability of ${}^nM_m$ on evidence $H\&{}^kK_g$ is also $1-{}^n\rho_m$. Consequently, the probability of ${}^nL_m$ on evidence $H\&{}^kK_g$ is the same as on evidence H alone.

Thus ${}^nL_m$ is independent of ${}^kK_g$ in H.

5. Any positive product is independent of any negative product in H, supposing the two products do not contradict one another.

Proof: As we have seen, the probability of ${}^nK_m$ on evidence H is ${}^n\pi_m$. If ${}^kL_g$ and ${}^nK_m$ do not contradict one another, it follows from 4 that the probability of ${}^kL_g$ on evidence $H\&{}^nK_m$ is the same as on evidence H alone, *viz.* ${}^k\rho_g$. It follows from the

Multiplication Axiom that the probability of $^nK_m \& {}^kL_g$ on evidence H is $^n\pi_m \cdot {}^k\rho_g$. But $^k\rho_g$ is not 0. It follows from the first converse ($T2$) of the Multiplication Axiom that the probability of ^nK_m on evidence $H \& {}^kL_g$ is $^n\pi_m$. Consequently, the probability of ^nK_m on evidence $H \& {}^kL_g$ is the same as on evidence H alone.

Thus ^nK_m is independent of ^kL_g in H.

6. Any positive product is independent of any consistent mixed product in H, supposing that the products have no common constituent and do not contradict one another.

Proof: If ^nK_m and ^jK_i have no common constituent, it follows from 1 that the probability of ^nK_m on evidence $H \& {}^jK_i$ is the same as on evidence H alone, viz., $^n\pi_m$. $^nK_m \& {}^jK_i$ is a positive product of $m+i$ A-properties. If ^kL_g does not contradict $^nK_m \& {}^jK_i$, it follows from 4 that the probability of ^kL_g on evidence $H \& {}^nK_m \& {}^jK_i$ is the same as on evidence H alone, viz., $^k\rho_g$. It follows from the Multiplication Axiom that the probability of $^nK_m \& {}^kL_g$ on evidence $H \& {}^jK_i$ is $^n\pi_m \cdot {}^k\rho_g$. But $^k\rho_g$ is not 0. It follows from the first converse of the Multiplication Axiom that the probability of ^nK_m on evidence $H \& {}^jK_i \& {}^kL_g$ is $^n\pi_m$. Consequently, the probability of ^nK_m on evidence $H \& {}^jK_i \& {}^kL_g$ is the same as on evidence H alone.

Thus ^nK_m is independent of $^jK_i \& {}^kL_g$ in H.

7. Any consistent mixed product is independent of any positive product in H, supposing that the products have no common constituent and do not contradict one another.

The proof of 7 with the aid of 6 is analogous to the proof of 5 with the aid of 4.

8. Any negative product is independent of any consistent mixed product in H, supposing that the two products have no common constituent and do not contradict one another.

Proof: If ^nL_m and ^jL_i have no common constituent, it follows from 3 that the probability of ^nL_m on evidence $H \& {}^jL_i$ is the same as on evidence H alone, viz., $^n\rho_m$. $^nL_m \& {}^jL_i$ is a negative product of $m+i$ A-properties. If ^kK_g does not contradict $^nL_m \& {}^jL_i$, it follows from 5 that the probability of ^kK_g on evidence $H \& {}^nL_m \& {}^jL_i$ is the same as on evidence H alone, viz., $^k\pi_g$. It follows from the Multiplication Axiom that the probability of $^nL_m \& {}^kK_g$ on evidence $H \& {}^jL_i$ is $^n\rho_m \cdot {}^k\pi_g$. But it follows from v that $^k\pi_g$ is greater than 0. It follows from the first converse of the Multiplication Axiom that the probability of ^nL_m on evidence $H\&^jL_i\&^kK_g$ is $^n\rho_m$. Consequently, the probability of ^nL_m on evidence $H\&^jL_i\&^kK_g$ is the same as on evidence H alone.

Thus ^nL_m is independent of $^jL_i\&^kK_g$ in H.

9. Any consistent mixed product is independent of any negative product in H, supposing that the products have no common constituent and that they do not contradict one another.

The proof of *9* with the aid of *8* is analogous to the proof of *7* with the aid of *6*.

10. Any two consistent mixed products are mutually independent in H, supposing that the two products have no common constituent and that they do not contradict one another.

Proof: It follows from *6* that the probability of ^nK_m on evidence $H\&^jK_i\&^vL_u$ is $^n\pi_m$ and from *8* that the probability of ^kL_g on evidence $H\&^nK_m\&^jK_i\&^vL_u$ is $^k\rho_g$. It follows from the Multiplication Axiom that the probability of $^nK_m\&^kL_g$ on evidence $H\&^jK_i\&^vL_u$ is $^n\pi_m.^k\rho_g$. On the other hand, the probability of ^nK_m on evidence H is $^n\pi_m$. It follows from *4* that the probability of ^kL_g on evidence $H\&^nK_m$ is $^k\rho_g$. It follows from the Multiplication Axiom that the probability of $^nK_m\&^kL_g$ on evidence H is $^n\pi_m.^k\rho_g$. Consequently, the probability of $^nK_m\&^kL_g$ on evidence $H\&^jK_i\&^vL_u$ is the same as on evidence H alone.

Thus $^nK_m\&^kL_g$ is independent of $^jK_i\&^vL_u$ in H.

In a similar manner we prove that $^jK_i\&^vL_u$ is independent of $^nK_m\&^kL_g$ in H.

Thus $^nK_m\&^kL_g$ and $^jK_i\&^vL_u$ are mutually independent in H.

It is seen that if the sequence $A_1, \ldots, A_n, \ldots$ constitutes a Normal Independence-Realm in H, then the probability on evidence H of any positive, negative, or consistent mixed product of A-properties equals the product of the probabilities on evidence H of the A-properties in question themselves. Thus in a Normal Independence Realm the probability of any product of properties can be computed according to the so-called Special Multiplication Theorem.

15. The Direct Principles of Maximum Probability and of Great Numbers

The sequence $A_1, \ldots, A_n, \ldots$ of properties is said to constitute a Bernoullian Independence-Realm in H, if, and only if, in addition to conditions ii and iii and iv and v of the preceding paragraph the following condition is also fulfilled:

vi. All members of the sequence have the same probability on evidence H. In symbols:

(6) $(i)(p_i = p)$

The Bernoullian Independence-Realms are thus Normal Independence-Realms of particularly simple structure. Normal Independence-Realms which are not Bernoullian might also be called Poissonian Independence-Realms.

The treatment will henceforth be confined exclusively to Bernoullian Independence-Realms.

For the conjunction of the symbolic expressions (2)–(5) of the preceding paragraph we shall use the abbreviation I, and for the conjunction of (2)–(6) the abbreviation I_b.

In a Bernoullian Independence Realm ${}^n\pi_m$ equals p^m and ${}^n\rho_m$ equals $(1-p)^m$.

On I_b as datum we can use the Multiplication Principle for computing the probability on evidence H of a consistent mixed product with m positive and $n-m$ negative constituents. The probability is $p^m.(1-p)^{n-m}$. Thus we have the following theorem:

$T29$ $I_b \rightarrow (Exc(k, m, g, n-m) \rightarrow P({}^k K_m \&^g L_{n-m}, H, p^m.(1-p)^{n-m}))$.

On I_b as datum we can further compute the probability on evidence H of the sum of all mixed products which can be formed of n given members of the A-sequence, when m of them are taken positively and the remaining $n-m$ negatively. We first use the Multiplication Principle for computing the probability of the products ($T29$). Thereafter we use the Special Addition Principle for computing the probability of their sum. For the sum in question we have the name ${}^i_m Q^n \&_{n-m}{}^i R^n$. (Cf. above p. 59.) The probability is $\binom{n}{m}.p^m.(1-p)^{n-m}$. Thus we have the following theorem:

$T30$ $I_b \rightarrow P({}^i_m Q^n \&_{n-m}{}^i R^n, H, \binom{n}{m}.p^m.(1-p)^{n-m})$.

The value $\binom{n}{m}.p^m.(1-p)^{n-m}$ is the greater the less the ratio $m:n$ differs from p^1. Considering the meaning of $\binom{n}{m}.p^m.(1-p)^{n-m}$ as a probability, we can express this arithmetical truth as follows:

The probability, in a Bernoullian Independence Realm, that any m of n given properties will be present and the rest absent in a thing is the greater, the less the ratio m : n differs from the probability p of the individual properties themselves.

This we shall call the Direct Principle of Maximum Probability (for Bernoullian Independence-Realms).

On I_b as datum we can, finally, compute the probability on evidence H of the sum of all mixed products which can be formed of n given members of the A-sequence, when first m_1 of them are taken positively and the remaining $n-m_1$ negatively, then m_2 of them positively and the remaining $n-m_2$ negatively, . . ., and finally m_k of them positively and the remaining $n-m_k$ negatively. We first use the Multiplication Principle for computing the probability of the products ($T29$). Thereafter we use the Special Addition Principle for computing the probability of the sums of products ($T30$). Lastly we use the Special Addition Theorem again for computing the probability of the sum of sums. For the sum of sums we have the name

$\overset{i}{\triangle}{}^{n}_{m_1,\ldots,m_k}$. (Cf. above p. 59.) The probability is $\sum\limits_{\mu=1}^{k} \binom{n}{m_\mu}.p^{m_\mu}.(1-p)^{n-m_\mu}$. Thus we have the following theorem:

$$T31 \quad I_b \to P(\ \overset{i}{\underset{m_1,\ldots,m_k}{\triangle}}{}^{n}\ ,\ H,\ \sum_{\mu=1}^{k} \binom{n}{m_\mu}.p^{m_\mu}.(1-p)^{n-m_\mu}).$$

If $1, \ldots, k$ are those values of μ for which the ratio $m_\mu : n$ falls in the interval ε round p, then the value $\sum\limits_{\mu=1}^{k} \binom{n}{m_\mu}.p^{m_\mu}.(1-p)^{n-m_\mu}$ approaches 1 as a limit as n is indefinitely increased.[1] Or, considering its meaning as a probability:

If $1, \ldots, k$ are those values of μ for which the ratio $m_\mu : n$ differs less than a given amount ε from the probability p of the individual properties, then the probability, in a Bernoullian Independence-Realm, that any m_1

---

[1] This statement has the character of what we have called (p. 180) a 'subsidiary datum.' The proof will be found in any textbook on probability. Since the proof does not involve any applications of the principles of probability, it will be of no interest to reproduce it here.

or . . . or m_k of n given properties will be present and the rest absent in a thing increases with n and approaches the maximum probability 1 as a limit.

This we shall call the Direct Principle of Great Numbers (for Bernoullian Independence-Realms).

The Direct Principles of Maximum Probability and Great Numbers for Normal Independence-Realms, in which the probabilities of the individual properties are equal, were first proved by James Bernoulli (1713).

A generalization of the Direct Principles of Maximum Probability and Great Numbers for all Normal Independence-Realms was first proved by Poisson (1832).

16. The Inverse Principles of Maximum Probability and of Great Numbers

Let there be v_1 fields of measurement ${}^1H_1, \ldots, {}^1H_{v_1}$. For their sum we introduce the name T_1.

Let there be v_2 fields of measurement ${}^2H_1, \ldots, {}^2H_{v_2}$. For their sum we introduce the name T_2.

Similarly, let there be $v_3, \ldots, v_w$ fields of measurement. For their sums we introduce the names $T_3, \ldots, T_w$.

For the sum of all the $v_1 + \ldots + v_w$ fields of measurement, *i.e.*, for ΣT_w, we introduce the name T. T is thus the name of a property, *viz.*, the property of being a member of some of the $v_1 + \ldots + v_w$ fields of measurement.[1]

Let there be a sequence of properties $A_1, \ldots, A_n, \ldots$.

Let this sequence of properties constitute a Bernoullian Independence-Realm in every one of the $v_1 + \ldots + v_w$ fields of measurement. Thus we have v_1 data ${}^1I_{b_1}, \ldots, {}^1I_{b_{v_1}}$ and $\ldots$ and v_w data ${}^wI_{b_1}, \ldots, {}^wI_{b_{v_w}}$. For the conjunction of all those data we shall use the abbreviation $\mathcal{J}_b$.

There are four more data of the problem to be treated, *viz.*:

---

[1] Instead of T we might also have used a symbol, say $T_{v_1, \ldots, v_w}$, retaining the indices $v_1, \ldots, v_w$. This complication, however, is not essential.

i. Any two of the $v_1 + \ldots + v_w$ fields of measurement are mutually exclusive. In symbols:

(1) $(g)(k)(m)(n)(k \leqslant w \ \& \ n \leqslant w \ \& \ g \leqslant v_k \ \& \ m \leqslant v_n \rightarrow ((E \ ^kH_g \& ^nH_m) \rightarrow k = n \ \& \ g = m))$.

ii. Any one of the properties $A_1, \ldots, A_n, \ldots$ has the same probability p_1 on any one of the v_1 properties $^1H_1, \ldots,$ $^1H_{v_1}$ as evidence. Any one of the properties $A_1, \ldots,$ $A_n, \ldots$ has the same probability p_2 on any one of the v_2 properties $^2H_1, \ldots, \ ^2H_{v_2}$ as evidence. And so on up to p_w and v_w. In symbols:

(2) $(i)(m)(n)(n \leqslant w \ \& \ m \leqslant v_n \rightarrow P(A_i, \ ^nH_m, \ p_n))$.

iii. The w probabilities $p_1, \ldots, \ p_w$ are all different. In symbols:

(3) $(m)(n)(m \leqslant w \ \& \ n \leqslant w \rightarrow (p_m = p_n \rightarrow m = n))$.

iv. The sum T_1 of the v_1 first fields of measurement has the probability q_1 on the sum T of all the $v_1 + \ldots + v_w$ fields of measurement as evidence. The sum T_2 of the v_2 next fields of measurement has the probability q_2 on the sum T of all the $v_1 + \ldots + v_w$ fields of measurement as evidence. And so on up to v_w and q_w. In symbols:

(4) $(n)(n \leqslant w \rightarrow P(T_n, \ T, \ q_n))$.

For the conjunction of the four symbolic expressions (1)–(4) we shall use the abbreviation U.

On $\mathcal{J}_b$ and U as data we can answer the following question:

What is the probability that a thing will belong to some of v_i mutually exclusive fields of measurement, in which the probability of any one of the properties $A_1, \ldots, A_n, \ldots$ is p_i, on the evidence:

α. that the thing belongs to some of $v_1 + \ldots + v_w$ mutually exclusive fields of measurement ($i \leqslant w$), in which the probabilities of any one of the properties $A_1, \ldots, A_n, \ldots$ are p_1 and $\ldots$ and p_w respectively, and

β. that of given n ones of the properties $A_1, \ldots, A_n, \ldots$ m are present and the remaining n-m absent in the thing?

Or, to state the same question more briefly: what is the probability of T_i on evidence $T \& {}^k K_m \& {}^g L_{n-m}$?

This question we shall call the Problem of Bayes. The q-values we shall call the probabilities *a priori*, and the computed values we shall call the probabilities *a posteriori* of the problem.

According to $T29$, the probability of ${}^k K_m \& {}^g L_{n-m}$ on ${}^1 H_1$ as evidence is $p_1{}^m.(1-p_1)^{n-m}$. Similarly, the probability of ${}^k K_m \& {}^g L_{n-m}$ on ${}^1 H_2$ as evidence is $p_1{}^m.(1-p_1)^{n-m}$. And so on up to ${}^1 H_{v_1}$.

Since the fields of measurement are mutually exclusive, it follows from $T28$, or the Special Composition Theorem, that the probability of ${}^k K_m \& {}^g L_{n-m}$ on T_1 as evidence is also $p_1{}^m.(1-p_1)^{n-m}$. Similarly, the probability of ${}^k K_m \& {}^g L_{n-m}$ on T_2 as evidence is $p_2{}^m.(1-p_2)^{n-m}$. And so on up to T_w.

The conjunction of a disjunction with one of its members is identical with that member. Thus $T \& T_1$ is identical with T_1. Consequently, the probability of ${}^k K_m \& {}^g L_{n-m}$ on $T \& T_1$ as evidence is also $p_1{}^m.(1-p_1)^{n-m}$. Similarly, the probability of ${}^k K_m \& {}^g L_{n-m}$ on $T \& T_2$ as evidence is $p_2{}^m.(1-p_2)^{n-m}$. And so on up to $T \& T_w$.

Thus for all values of i from 1 to w the probability of T_i on evidence T is q_i and the probability of ${}^k K_m \& {}^g L_{n-m}$ on evidence $T \& T_i$ is $p_i{}^m.(1-p_i)^{n-m}$. From $T26$ or the Inverse Principle it then follows that the probability of T_i on evidence $T \& {}^k K_m \& {}^g L_{n-m}$ is $p_i{}^m.(1-p_i)^{n-m}.q_i : \sum\limits_{\mu=1}^{w} p_\mu{}^m.(1-p_\mu)^{n-m}.q_\mu$. Thus we have proved the theorem:

$T32$ $\mathcal{J}_b \& U \rightarrow (Exc(k, m, g, n-m) \& i \leq w \rightarrow$
$P(T_i,\ T \& {}^k K_m \& {}^g L_{n-m},\ p_i{}^m.(1-p_i)^{n-m}.q_i : \sum\limits_{\mu=1}^{w} p_\mu{}^m.(1-p_\mu)^{n-m}.q_\mu))$.

Traditionally particular importance has been attached to the case, where there is still one more datum, *viz.*

v. All the probabilities *a priori* are equal. In symbols:

(5) $(n)(n \leq w \rightarrow q_n = q)$.

(It might be noted that q equals $1 : w$. For, according to $T24'$, or the Special Addition Theorem, the probability of ΣT_w on evidence T is Σq_w. But ΣT_w is identical with T. Thus the probability of ΣT_w on evidence T is, according to the Inclusion Axiom, also 1. From the Axiom of Uniqueness it follows that $1 = \Sigma q_w$. If all q-values equal the value q, we have $1 = qw$ or $q = 1 : w$.)

For the conjunction of the five symbolic expressions (1)-(5) we may use the abbreviation U_b.

On the data $\mathcal{J}_b$ and U_b the probabilities *a priori* do not influence the calculated probabilities *a posteriori*. We therefore get the simplified theorem:

$T33 \quad \mathcal{J}_b \& U_b \rightarrow (Exc(k, m, g, n\text{-}m) \& i \leqq w \rightarrow$

$$P(T_i, \; T \& {}^k K_m \& {}^g L_{n\text{-}m}, \; p_i{}^m.(1\text{-}p_i)^{n\text{-}m} : \sum_{\mu=1}^{w} p_\mu{}^m.(1\text{-}p_\mu)^{n\text{-}m})).$$

The numerator $p_i{}^m.(1\text{-}p_i)^{n\text{-}m}$ is the greater, the less the ratio $m:n$ differs from p_i.[1] Hence also the quotient $p_i{}^m.(1\text{-}p_i)^{n\text{-}m} : \sum_{\mu=1}^{w} p_\mu{}^m.(1\text{-}p_\mu)^{n\text{-}m}$. Or, considering its meaning as a probability:

If, in the Problem of Bayes, the probabilities a priori *are equal, then the probability* a posteriori *that a thing in which m of n given properties are present and the rest absent, will belong to a field of measurement, in which the probability of the individual properties is p_i, is the greater, the less the ratio m : n differs from p_i.*

This we shall call the Inverse Principle of Maximum Probability (for Bernoullian Independence-Realms).

On $\mathcal{J}_b$ and U as data we can also answer the following question:

What is the probability that a thing will belong to some of v_{i_1} mutually exclusive fields of measurement, in which the probability of any one of the properties $A_1, \ldots, A_n, \ldots$ is p_{i_1}, or ... or to some of v_{i_j} mutually exclusive fields of measurement, in which the probability of any one of the properties $A_1, \ldots, A_n, \ldots$ is p_{i_j}, on the evidence:

a. that the thing belongs to some of $v_1 + \ldots + v_w$ mutually exclusive fields of measurement (i_1 and ... and i_j not being greater than w), in which the probabilities of any one of the properties $A_1, \ldots, A_n, \ldots$ are p_1 and ... and p_w respectively, and

β. that of given n ones of the properties $A_1, \ldots, A_n, \ldots$ m are present and the remaining n-m ones absent in the thing?

Or, to state the question more briefly: what is the probability of ΣT_{i_j} on evidence $T \& {}^k K_m \& {}^g L_{n\text{-}m}$?

Since the members of the sum ΣT_{i_j} are mutually exclusive,

[1] Cf. footnote on p. 205.

an application of $T27$ or the Extended Inverse Theorem gives
the value $\sum\limits_{\mu=1}^{j} p_{i_\mu}{}^m.(1-p_{i_\mu})^{n-m}.q_{i_\mu} : \sum\limits_{\mu=1}^{w} p_\mu{}^m.(1-p_\mu)^{n-m}.q_\mu.$

Thus we have the following theorem:

$T34 \quad \mathcal{J}_b \& U \to (Exc(k, m, g, n-m) \& (u)(u \leqslant j \to i_u \leqslant w) \to$

$\quad P(T_{i_j}, \, T \& {}^k K_m \& {}^g L_{n-m}, \, \sum\limits_{\mu=1}^{j} p_{i_\mu}{}^m.(1-p_{i_\mu})^{n-m}.q_{i_\mu} : \sum\limits_{\mu=1}^{w} p_\mu{}^m.(1-p_\mu)^{n-m}.q_\mu)).$

If $p_{i_1}, \ldots, p_{i_j}$ are those p-values which fall in the interval ε
round the ratio $m : n$, and if at least one of the corresponding
q-values $q_{i_1}, \ldots, q_{i_j}$ is greater than o, then the value $\sum\limits_{\mu=1}^{j} p_{i_\mu}{}^m.$
$(1-p_{i_\mu})^{n-m}.q_{i_\mu} : \sum\limits_{\mu=1}^{w} p_\mu{}^m.(1-p_\mu)^{n-m}.q_\mu$ approaches 1 as a limit as n is
indefinitely increased.[1] Or, considering probability:

*If, in the Problem of Bayes, $p_{i_1}, \ldots, p_{i_j}$ are those probabilities
of the individual properties which fall in the interval ε round the ratio
$m : n$; and if not all the corresponding probabilities a priori are o; then
the probability a posteriori that a thing, in which m of n given properties are present and the rest absent, will belong to a field of measurement, in which the probability of the individual properties differs less
than the given amount ε from the ratio $m : n$ increases with n and
approaches the maximum probability 1 as a limit.*

This we shall call the Inverse Principle of Great Numbers
(for Bernoullian Independence-Realms).

*　　*　　*　　*　　*

In text-books on probability, the Problem of Bayes is usually
treated on the suppositions that w is non-denumerably great
and that the p-values continuously cover the whole range from
o to 1. A treatment on these suppositions is, however, beyond
the reach of the Elementary Calculus of Probability.

17. Pseudo-Deductions of the Inverse Principles of Maximum Probability and of Great Numbers

The so-called probability *a posteriori* of the Problem of Bayes
is thus the probability of a certain field of measurement of

---

[1] Cf. footnote on p. 205.

another probability. This 'superimposition of probabilities' is frequently misrepresented.

It is, *e.g.*, sometimes said that the probability *a posteriori* of the Problem of Bayes is the probability of a probability. This mode of expression suggests the idea that there is some sort of distinction of logical types involved in the problem. The probability *a posteriori* seems to refer to a property of higher type than the properties to which the p-values refer.

It is reasonable to think that the idea of a type difference between the probability *a posteriori* and the p-values has its root in a false conception of the nature of the probabilities *a priori*. The crucial mistake would consist in the fact that the above T-fields of measurement are thought of as *sets* and not as *sums* (disjunctions) of the underlying H-fields of measurement. This fallacy is characteristic, *e.g.*, of the deduction of the Inverse Principle of Great Numbers given by Reichenbach in his *Wahrscheinlichkeitslehre* (§§56–60).

The difference between the two conceptions of the probabilities *a priori*, the sum-conception leading to a correct deduction of the principles, and the set-conception providing a pseudo-deduction of them, will be illustrated by means of a concrete example.

Let us imagine an urn containing a large number of small dice. The dice are not all homogeneous; some of them are 'correct' but others are biased. As a result, say, of experience from a long series of throws we believe ourselves to know the probabilities of getting a one with each one of the dice. Thus there are in the urn v_1 dice with the probability p_1 for getting a one, v_2 dice with the probability p_2 of getting a one, and finally v_w dice with the probability p_w of getting a one. The properties $A_1, \ldots, A_n, \ldots$ mean: a one in the first throw, $\ldots$, a one in the n:th throw, $\ldots$. The results in throwing are, as is known, regarded as independent of one another. Thus the conditions for the presence of a Bernoullian Independence-Realm are taken to be satisfied.

The Problem of Bayes is now this:

A die is drawn from the urn. With this die n throws are made. Of them m are ones and the rest not. What is the probability that the drawn die was one with which the probability of getting a one is p_i?

If the probability of getting a one with the drawn die is p_i, then the probability of getting m ones in n tosses would be $p_i^m . (1-p_i)^{n-m}$. In order to answer the above question, however, it is necessary also to know the probabilities *a priori* of, as we say, the different probabilities of getting a one.

Let ϕ denote the property of being a die from the urn. Let ϕ_1 denote the property of being a die with which the probability of getting a one is p_1. Similarly, we introduce ϕ_2, and so on up to ϕ_w. The probabilities *a priori* of the Problem of Bayes, one might suggest, are the probabilities of ϕ_i on evidence ϕ, or else the probabilities that a die will have the probability p_i for a one. (i runs through all values from 1 to w.) These probabilities might, *e.g.*, be interpreted as the quotients $v_1 : (v_1 + \ldots + v_w)$ or the relative frequencies of *dice* in the urn with the probability p_i for a one.[1]

However, this way of viewing the required probabilities *a priori* is wrong. The right way of proceeding is the following:

Let T denote the property of being a throw with a die from the urn. (Cf. above p. 206.) Let T_1 denote the property of being a throw with some of the v_1 dice, with which the probability of getting a one is p_1. Similarly, we introduce T_2, and so on up to T_w. (Cf. above p. 206.) The required probabilities *a priori* of the Problem of Bayes are the probabilities of T_i on evidence T, or else the probabilities that a throw will be with a die, of which the probability p_i for a one is characteristic. (i runs through all values from 1 to w.) These probabilities might, *e.g.*, be interpreted as the relative frequencies of *throws* with dice in the urn with the probability p_i for a one.

It is evident that the two ways, the formally correct and the formally incorrect one, of forming the probabilities *a priori may* 'materially' lead to concordant results, *i.e.*, give the same values. But it is also evident that such a concordance would be purely contingent and not necessary. If the probabilities are interpreted as frequencies in the way suggested above, then concordance would mean that the distribution of the various probabilities of getting a one among the *drawn dice* gives a true picture of the distribution of those probabilities among the *dice in the urn*. It is reasonable to expect that this would be the case, if the urn were symmetrically built and did not contain parts where the dice can 'hide themselves,' if the dice are all of equal size and weight (though not equally balanced), and if, finally, the contents of the urn were carefully mixed before drawing. In imagining the urn, we tend to imagine it under these 'ideal' conditions, but it is plain that they are purely accidental to the problem as such.

[1] This interpretation corresponds to what Reichenbach (*op. cit.*, §58) calls 'vertikale Abzählung im Wahrscheinlichkeitsgitter.'

18. The Principle of Succession

On the same premises as in the Problem of Bayes, *i.e.*, on the data which we have called J_b and U respectively, we can finally answer the following question:

What is the probability that a thing will have the property A_{n+1} on the evidence:

a. that it belongs to some of $v_1 + \ldots + v_w$ mutually exclusive fields of measurement, in which the probabilities of any one of the properties $A_1, \ldots, A_n, \ldots$ are p_1 and $\ldots$ and p_w respectively, and

β. that it is a positive instance of the first n ones of the properties $A_1, \ldots, A_n, \ldots$?

Or, using the same symbols as in the Problem of Bayes: what is the probability of A_{n+1} on evidence $T \& \Pi A_n$?

The question can also be raised and answered for the more general case, when the thing is a positive instance of m of some n properties, and a negative instance of the remaining $n-m$ ones. This case, however, will not concern us in this inquiry.

Since the properties $A_1, \ldots, A_n, \ldots$ are totally independent in ${}^1H_1$, the probability of A_{n+1} on evidence ${}^1H_1 \& \Pi A_n$ is the same as on evidence ${}^1H_1$ alone, *viz.* p_1.

Similarly, the probability of A_{n+1} on evidence ${}^1H_2 \& \Pi A_n$ is p_1, and $\ldots$, and the probability of A_{n+1} on evidence ${}^1H_{v_1} \& \Pi A_n$ is p_1.

Thus, according to *T28* or the Special Composition Theorem, the probability of A_{n+1} on evidence ${}^1H_1 \& \Pi A_n \text{v} \ldots \text{v} {}^1H_{v_1} \& \Pi A_n$ is also p_1.

${}^1H_1 \& \Pi A_n \text{v} \ldots \text{v} {}^1H_{v_1} \& \Pi A_n$ is identical with $T_1 \& \Pi A_n$. (Cf. above p. 192.)

$T_1 \& \Pi A_n$ is identical with $T \& T_1 \& \Pi A_n$. (Cf. above p. 208.)

According to *T32*, the probability of T_1 on evidence $T \& \Pi A_n$ is $p_1^n . q_1 : \sum\limits_{\mu=1}^{w} p_\mu^n . q_\mu$.

Thus, according to the Multiplication Axiom, the probability of $T_1 \& A_{n+1}$ on evidence $T \& \Pi A_n$ is $p_1^{n+1} . q_1 : \sum\limits_{\mu=1}^{w} p_\mu^n . q_\mu$.

Similarly, the probability of $T_2 \& A_{n+1}$ on evidence $T \& \Pi A_n$ is

$p_2{}^{n+1} . q_2 : \sum\limits_{\mu=1}^{w} p_\mu{}^n . q_\mu$, and . . ., and the probability of $T_w \& \Pi A_{n+1}$

on evidence $T \& \Pi A_n$ is $p_w{}^{n+1} . q_w : \sum\limits_{\mu=1}^{w} p_\mu{}^n . q_\mu$.

Since the products $T_1 \& A_{n+1}, \ldots, T_w \& A_{n+1}$ are mutually exclusive, it follows from $T24'$ or the Special Addition Theorem that the probability of $T_1 \& A_{n+1} \text{v} \ldots \text{v} T_w \& A_{n+1}$ on evidence

$T \& \Pi A_n$ is $\sum\limits_{\mu=1}^{w} p_\mu{}^{n+1} . q_\mu : \sum\limits_{\mu=1}^{w} p_\mu{}^n . q_\mu$.

The sum $T_1 \& A_{n+1} \text{v} \ldots \text{v} T_w \& A_{n+1}$ is identical with $T \& A_{n+1}$.

According to $T8$ or the Theorem of Equivalence, the probability of $T \& A_{n+1}$ on evidence $T \& \Pi A_n$ is the same as the probability of A_{n+1} alone on evidence $T \& \Pi A_n$.

Thus the probability of A_{n+1} on evidence $T \& \Pi A_n$ is $\sum\limits_{\mu=1}^{w} p_\mu{}^{n+1} . q_\mu : \sum\limits_{\mu=1}^{w} p_\mu{}^n . q_\mu$.

We have thus proved the theorem

$T35 \quad \mathfrak{J}_b \& U \rightarrow P(A_{n+1}, \; T \& \Pi A_n, \; \sum\limits_{\mu=1}^{w} p_\mu{}^{n+1} . q_\mu : \sum\limits_{\mu=1}^{w} p_\mu{}^n . q_\mu)$.

If at least one of the q-values is greater than o, the ratio $\sum\limits_{\mu=1}^{w} p_\mu{}^{n+1} . q_\mu : \sum\limits_{\mu=1}^{w} p_\mu{}^n . q_\mu$ increases with n. Or, considering probability:

If, in the Problem of Bayes, not all the probabilities a priori are o, then the probability that a thing which is a positive instance of the n first members of a sequence of properties, will also be a positive instance of the next member of the sequence, increases with n.

(The increasing probability approaches 1 as its limit, if there is a field of measurement which is not *a priori* minimally probable, and in which the probability of any one of the properties $A_1, \ldots, A_n, \ldots$ is 1.)

This we shall call the Principle of Succession.

*   *   *   *   *

Under the two simplifying assumptions that the p-values continuously cover the range from o to 1, and that all the q-values are equal, we get for the probability in question the famous value $\int\limits_0^1 p^{n+1} \, dp : \int\limits_0^1 p^n \, dp$ or $(n+1) : (n+2)$. The deduction of this answer to our question, is, however, beyond the reach of the Elementary Calculus of Probability.[1]

*   *   *   *   *

The Inverse Principles in general and those of Maximum Probability, Great Numbers, and Succession in particular, have sometimes been thought to be of the greatest importance to the problems of induction. This opinion will be examined in Chapter X. It will be shown to be largely an illusion. It is a plausible suggestion that the misconceptions concerning the applicability of the Inverse Principles for purposes of induction have been due to an insufficient insight into the fairly complicated logical mechanism of those principles.

19. The Interpretation of the Calculus and the Analysis of Probability

The axiomatic and formalized deductive system which we have outlined in §§2–18 of this chapter contains, besides familiar notions of logic and arithmetic, one other notion, *viz.* the relation P which we introduced in §1.

It is characteristic of the system that it does not give us any means of constructing, for any given formula in which the name of this new concept occurs, another formula without it which is identical with the first. The system, in other words, does not include an explicit definition of the notion of probability in terms of other notions.

If, in the formulae of the calculus, we replace P by the name of another non-homogeneous relation, any order of which is also an order of P, then we get an interpretation of the calculus. (This description is somewhat inaccurate, but will suffice for our purposes.)

If, on an interpretation of the calculus, the formulae express true propositions, we speak of a true or valid interpretation. If, on an interpretation, the formulae express formal propositions

---

[1] It might be mentioned that the value $(n+1) : (n+2)$ can be obtained without integration as an answer to the following question: What is *the limiting value* which the calculated probability in our problem approaches, when w is indefinitely increased, on the two simplifying assumptions that the p-values are $p_1=0$, $p_2=1 : (w-1)$, $p_3=2 : (w-1)$, . . ., $pw=1$ and that all the q-values are equal? This was pointed out to me by Professor C. D. Broad.—There are also other 'special cases', in which the value can be got without integration.

we talk of a formal, and if they express material propositions we talk of a material, interpretation of the calculus.

From the problem of interpreting the calculus we must distinguish the problem of analysing the relation *P*.

The meaning of (logical) analysis is vague and complex. Analyses may answer to many different purposes and may differ in nature accordingly. *One* purpose of analysis has to do with verification or the process of coming to know the truth-value of propositions. With *that* purpose in mind, we would call something an analysis of probability which facilitates the deciding of the truth-value of probability-propositions. The meaning of 'facilitate' is vague and we shall not here attempt to make it clear.

It should be observed that an interpretation of the calculus may make it possible for us to decide the truth-value of the propositions expressed by the formulae (axioms and theorems), without providing a means of deciding the truth-value of the propositions expressed by the probability-expressions themselves. Of this we shall soon give an example. Therefore, a valid interpretation of the calculus need not be an adequate analysis (in the above or some other sense) of probability.

It should further be observed that an interpretation of the calculus may make the formulae (axioms and theorems) express formal propositions and the probability-expressions express material propositions. Of this also we shall soon give an example.

20. *Frequency, Possibility, Degree of Belief, and Probability*

The symbol $P(A, H, p)$ can be interpreted as meaning that p is the proportion or relative frequency of things with the property H which also possess the property A. Generally speaking: the probability-value is the proportion of positive instances of the evidence-property which are also positive instances of the conjectured property.

The notion of a proportion, we said (p. 78), makes sense only on condition that the property of which a proportion is contemplated, is either finite or replaced by a sequence. It should be observed that if a certain evidence-property is replaced by a sequence, it must be replaced by *the same* sequence throughout the entire formula in which the probability-expression occurs. This is obvious since, by rearranging a sequence, we can alter a proportion. (Cf. above p. 80 f.)

In either case we get a valid formal interpretation of our Calculus of Probability.[1] We call this the Frequency Interpretation.

On the Frequency Interpretation the axioms and theorems of axiomatic probability express formal propositions. This should be taken as the precise content of our previous statement (p. 169), that the calculus can be deductively built up directly on the basis of the definition of probability as a frequency.

On the Frequency Interpretation the probability-propositions themselves are either formal or material propositions about proportions. *E.g.*, it is a formal proposition that the proportion of primes among cardinal numbers is 0, but it is a material proposition that the proportion of 'heads' among throws with a certain coin is 1 : 2. Material propositions about proportions in infinite (*i.e.*, numerically unrestricted) populations are Statistical Laws.

According to whether, on the Frequency Interpretation, probability-propositions are formal or material we may distinguish between formal or intensional, and material or extensional, probability.[2]

Take a formula (axiom or theorem) of the calculus. Replace the names of evidence-properties which occur in the formula by their perfect disjunctive normal denotations in terms of all atomic names of properties which occur in the formula. If the formula contains a sentence stating that a certain one of the properties denoted by a conjunction-name in some of the normal denotations is empty, then omit this conjunction-name from the normal denotation. The number of (remaining) conjunction-names in the respective normal denotations is the number of ways, as we say,

---

[1] The proof will not be given here.

[2] To say of a probability that it is, in our sense, 'extensional' is not an assertion about the way in which this probability is *actually* being interpreted. The probability 1:2 of getting 'heads' in tossing with a coin e.g. is 'extensional' independently of whether we interpret it as a statistical frequency or as a ratio of possibilities. It is 'extensional' because, *if* interpreted statistically, it cannot be decided on formal grounds.

in which the respective evidence-properties can exist with regard to the conjectured properties. The number of (remaining) conjunction-names in the respective normal denotations which entail the name of the respective conjectured properties is, as we say, the number of 'favourable' ways or the number of ways in which the respective conjectured properties and evidence-properties can co-exist.

The symbol $P(A, H, p)$ can be interpreted as meaning that p is the ratio of the number of possible ways in which A and H can co-exist, to the number of possible ways in which H can exist with regard to A. Generally speaking, the probability-value is the ratio of the number of ways in which the conjectured property and the evidence-property can co-exist, to the number of ways in which the evidence-property can exist with regard to the conjectured property.

An example will illustrate this. Take the formula $(E\ H)\rightarrow((H \subset A)\rightarrow P(A,\ H,\ 1))$. The only evidence property which occurs in the formula is the one denoted H. Its perfect disjunctive normal denotation in terms of all atomic names of properties which occur in the formula is $H\&AvH\&\overline{A}$. The sentence $H \subset A$ which occurs in the formula is identical with $\overline{E}H\&\overline{A}$. Thus the conjunction-name $H\&\overline{A}$ should be omitted from the normal denotation $H\&AvH\&\overline{A}$. What remains is $H\&A$. The number of ways in which the evidence-property can exist with regard to the conjectured property is thus 1. But $H\&A$ entails A. The number of ways in which the conjectured property and the evidence-property can co-exist is thus also 1. Hence the probability or the ratio of 'favourable' to possible cases is 1.

The interpretation which uses the above device we call the Possibility Interpretation. It can be proved to be a valid formal interpretation of our Calculus of Probability.[1]

On the Possibility Interpretation also, the axioms and theorems of axiomatic probability express formal propositions. (*E.g.*, that the probability is 1 if the evidence-property is included in the conjectured property means, on this interpretation, that the evidence-property cannot exist without co-existing with the conjectured property.) This should be taken

[1] The proof will not be given here. It should be observed that, on adding a further device, the interpretation is valid also for the Axiom of Continuity.

as the precise content of our previous statement (p. 168) that the calculus can be deductively built up directly on the basis of the definition of probability as a ratio of possibilities.

It is a difference between the Frequency and the Possibility Interpretation that on the second, the probability-expressions themselves without exception express formal propositions. *E.g.*, it is a formal proposition that, if a toss with a coin can be either 'heads,' or 'tails' but not both, then the ratio of the number of possibilities favourable to 'heads' to the total number of possibilities is 1 : 2, just as it is a formal proposition that, if something is a cardinal, then there is an 'infinitely small' possibility that it is a prime.

Finally, one might suggest an interpretation of the symbol $P(A, H, p)$ as meaning that p is the degree to which we believe in the presence of A in a thing on the evidence that H is present in the same thing. Generally speaking, the probability-value would be the degree of belief in the presence of the conjectured property which we entertain on knowing the presence of the evidence-property.

We call this the Psychological Interpretation.

As already observed (p. 169 f.), it is uncertain whether any satisfactory method for measuring partial belief (independent of probability) could be provided. For the sake of argument, however, we shall assume that there is such a method. This assumption would carry with it the following observations:

On the Psychological Interpretation the axioms and theorems of axiomatic probability would *not* express formal propositions but material propositions about the way in which people actually distribute their partial belief. They would be, so to speak, psychological laws of believing. The mere fact that the interpretation would make probability 'subjective' in the sense that different persons (or even the same person at different times) may entertain different degrees of belief in the same property on the same evidence, is, as such, no proof that probability is 'subjective' also in the further sense that not all people distribute their beliefs in accordance with the axioms and theorems of the calculus. Nevertheless, probability *may* be 'subjective' in this further sense also, *i.e.*, the axioms and theorems as laws of believing *may* be false. There are, moreover, certain well-known psychological phenomena, the so-called maturity-of-odds arguments, which seem to indicate that these laws *are* false, *i.e.*, that people do not without exception believe in conformity with the rules of the calculus of Probability. Whether this really is so or not cannot, of course, be decided until there is a reliable method for measuring partial belief. I mention the point, because it seems to me to be worth while to tackle the problem of finding out to what extent the rules of probability have a psychological significance.

There is thus in any case a fundamental difference between the two first and the third interpretation of axiomatic probability. The Frequency and the Possibility Interpretations make the axioms and theorems of the Calculus deducible from a definition of probability in virtue of laws of logic and arithmetic. The Psychological Interpretation, on the other hand, would not make these axioms and theorems deducible from a definition of probability, but would make them psychological laws of believing, to be confirmed or refuted by experience. In a sense, therefore, the Frequency and the Possibility Interpretations are analogous to the arithmetical interpretation of axiomatic Euclidean geometry, whereas the Psychological Interpretation would be analogous to the interpretation of axiomatic geometry as a theory of physical space. (Needless to say, there is no comparison whatsoever, as regards their importance, between geometry as a theory of space and the Calculus of Probability as a suggested psychological theory of belief.)

*   *   *   *   *

We have now shown in what sense there is or could be an *interpretation* of the calculus in terms of frequency, possibility, and partial belief. It remains to say something about the *analysis* of probability in terms of the three notions.

Suppose that we can decide whether the evidence-property and the conjectured property are present in a thing or not.

Suppose that the number of positive instances of the evidence-property is known to be (not greater than) n. Then it is possible to decide whether the proposition that a proportion p of the positive instances of the evidence-property are also positive instances of the conjectured property is true or false.

Suppose that the number of positive instances of the evidence-property is not known to be (not greater than) n. Suppose the proposition that the proportion p of the positive instances of the evidence-property are also positive instances of the conjectured property, to be a material proposition. Then it is not possible to decide whether this proposition is true or false on the basis of a decision for each positive instance of the evidence-property that it is or is not a positive instance of the conjectured property also.

We shall call a relative frequency in a population with an unknown (finite or infinite) number of members a long-run frequency. Limiting frequencies are thus a sub-species of long-run frequencies. It should be observed that the difficulties of verification and falsification of material propositions about probabilities arise as soon as we have to do with long-run frequencies.

The above observations will justify the following conclusions as regards the analysis of probability in terms of frequency:

In the case of extensional probability there is a frequency analysis, provided that it is possible to decide whether the conjectured and the evidence property are present in any given thing or not, *and* that the number of positive instances of the evidence-property is known to be (not greater than) n.

Only in the case of intensional probability can there be an analysis in terms of long-run or limiting frequency.

These conclusions are based on the fact that the interpretation of extensional probability as a long-run frequency does not in any way help us to come to know the truth-value of probability-propositions. We are just as ignorant after as we were before the interpretation, whether such propositions are true or false. To say that by the probability of getting 'heads' in tossing a coin we understand the limiting frequency of 'heads' is, from the point of view of verification, no more helpful or illuminating than it would be to say that by the limiting frequency of 'heads' we understand the probability of 'heads' in tossing. Those authors who insist upon a frequency analysis of probability throughout, seem to me to be victims of a tendency to assimilate *all* cases of probability to a pattern which provides an analysis in *some* cases. This assimilation is possible, since no findings of experience can ever contradict a proposition about a limiting frequency. But the assimilation ceases to be useful or illuminating so soon as it tries to explain extensional probability in terms of long-run frequency.

From the point of view of the analysis of probability the possibility definition enjoys the advantage over the frequency definition that its verification problem is affected neither by the number of positive instances of the evidence-property, nor by difficulties over deciding whether the conjectured property and the evidence-property are present or absent in a thing.

The limitations of the analysis in terms of possibility are clearly seen in cases in which we make an estimate of probability without being able to point to a corresponding ratio of possibilities. We all agree that in the case of a heavily biased coin the probability of getting 'heads' in tossing is not 1 : 2. But where are the possibilities in terms of which the true value of the probability is now to be analysed? (Cf. below p. 232 f.)

Whether an interpretation of probability in terms of partial belief would be helpful or not from the point of view of analysis depends in the first place upon whether a satisfactory method of measuring degrees of belief could be provided or not. The existence of such a method being debatable, we shall not discuss further the analytic value of a psychological definition of probability.

$$* \qquad * \qquad * \qquad * \qquad *$$

In this inquiry we shall not make any other claims on behalf of the different interpretations than that two of them, the Frequency Interpretation and the Possibility Interpretation, are valid formal interpretations of our Calculus of Probability. The question of the analysis of probability in the sense of the problem, how we may come to know the truth-value of probability-propositions, will not concern us further.

Chapter Eight

PROBABILITY AND PREDICTION

Induction and Probability

IN this and the next two chapters some aspects of the relation of probability to inductive inference will be studied.

Probability, at least when it is of the kind which we have (p. 217) called extensional, is used as a substitute for certainty, *i.e.*, as an attribute of propositions the truth-value of which is not known to us from past or present experience but is conjectured or anticipated. One main type of anticipated propositions are inductive conclusions. (Cf. Chap. I, §1.) If probability as an attribute of inductive conclusions is called Inductive Probability, it would follow that most cases of extensional probability are cases of Inductive Probability.

Such a comprehensive use of the term Inductive Probability is, however, not very convenient. We shall here reserve the term exclusively for probability as an attribute of inductive conclusions of the second order, *i.e.*, of theories and of the particularly important kind of theories which we call laws (Laws of Nature). Thus probability as an attribute of inductive conclusions of the first order, *i.e.*, of predictions, will not be called Inductive Probability.

It is usually not difficult to discern whether, in a given case, probability is being used as an attribute of an inductive conclusion of the first or of the second order. There are, however, some interesting cases which are liable to cause confusion. We shall therefore distinguish between Real Inductive Probability which is probability as a genuine attribute of theories or laws, and Apparent Inductive Probability which is really an attribute of predictions but apparently of theories or laws.

The most important cases of Apparent Inductive Probability occur in connexion with certain classical problems grouped under the heading of Inverse Probability. Their treatment involves applications of the theorems of the calculus which we have called the Inverse Principle, the Inverse Principle of Maximum Probability, and the Inverse Principle of Great Numbers. In these problems it may easily seem that we are dealing with the probability of a law when we are, in fact, dealing with the probability of a prediction. Traditionally, the term Inductive Probability has been used as almost synonymous with Inverse Probability. This peculiar usage is no doubt largely due to the traditional confusion over the logical nature of Inverse Probability. We shall discuss this topic in some detail in Chap. X.

Another type of cases of Apparent Inductive Probability has its root in a tendency to 'transfer' the probability of a prediction to the law of which the prediction in question is a test-condition. The probability p of the prediction that *a random* thing which is known to be H will also be A, is not infrequently confused with the probability of the law that all things which are H are also A, or $H \subset A$. This confusion, it seems, is particularly close at hand when there exists a high probability in favour of the occurrence of the conjectured property relative to the occurrence of the evidence property. The tendency to confuse a high probability in favour of the occurrence of A on evidence H with a high probability in favour of the law $H \subset A$ may perhaps be connected somehow with the fact that, in practice, we frequently do not acknowledge isolated disconfirming instances of $H \subset A$ as falsifications of the law. (Cf. Chap. VI, §§2 and 3.)[1]

* * * * *

In this chapter we shall deal exclusively with probability as an attribute of predictions.

[1] A remarkable instance of the above confusion is found in Reichenbach's paper *Über Induktion und Wahrscheinlichkeit* in Erkenntnis 5 (1935). The author here proposes that the probability of a law or Universal Implication $H \subset A$ could be measured by the proportion of true predictions of A from the presence of H. This proposal entails, *e.g.*, that if every second positive instance of H turns out to be A and every second not, then the probability that *all* H's will be A is 1 : 2. Reichenbach (*ibid.*, p. 274) seems prepared to accept this, though it is plainly repugnant to ordinary thinking and should rather be taken as a *reductio ad absurdum* of the proposal itself. Cf. Popper, *Logik der Forschung*, p. 191.

The topic is a large one. It may be divided into sections, according to the logical nature of the predictions concerned.

A most important part of the study has to deal with statistical predictions on the basis of samples. It belongs to that branch of the Logic of Induction which we have called the Logic of Statistical Inference. (Cf. Chap. III, §3.)

Another important section deals with (numerical) predictions on the basis of Quantitative Laws of Nature. Along with them may also be counted the study of 'errors.' It belongs to that branch of the Logic of Induction which we have called the Logic of Quantitative Induction. (Cf. Chap. III, §3.)

In conformity with our general limitation of the subject-matter of inquiry to the most general patterns of inductive inference (of the first and the second order), we shall not in this book deal with the logic of sampling or of errors, important as the topic is from the point of view of methodological study. Our treatment will be restricted to one problem only of a very general nature. We might call it the problem of chance.

2. *The Idea of Chance*

As regards the use of extensional probability the following distinction seems to be of some importance:

Suppose we conjecture about the occurrence of a property A in positive instances of a property H. Some instances may individually be more (or less) likely than other instances to have the feature A, *e.g.*, because they possess in addition to H a feature G also, such that the probability of A on evidence $H\&G$ is different from the probability of A on evidence H alone. Sometimes this fact does not concern us; what matters is merely the probability of A on evidence H. Sometimes, however, this fact *does* interest us and what matters is the probability of possessing the feature A which the instances have individually, *i.e.*, depending upon whether they do or do not exhibit certain other features in addition to H.

Of cases, where we are interested in individual variations in the probability of a property A within a main field of measurement H, one may further distinguish two principal types:

Sometimes our interest is confined to a more or less strictly defined set of (logically totally independent) properties ϕ_0. What matters is, which properties from ϕ_0 (or presence-functions of them) the instances possess in addition to H, such that the probability of A on H and (the product of) those properties as evidence, is different from the probability of A on H as sole evidence.

Sometimes, seemingly at least, our interest is not restricted to a limited range of properties only. What matters are *all* properties which the instances possess in addition to H, and which are such that the probability of A *and* (the product of) those properties as evidence, is different from the probability of A on evidence H.

In cases of the second type, however, the phrase '*all* properties' must be understood as subject to the restriction that the properties do not entail the presence or absence of the conjectured property A. For, if the evidence entails the presence of the conjectured property, its probability is 1, and if the evidence entails the absence of the conjectured property, its probability is 0, and these probabilities, moreover, are *intensional*, since the propositions that they are 1 and 0 respectively are formal propositions on the Frequency Interpretation.

Similarly, when we use the phrase 'no property' without mentioning ϕ_0 we always mean 'no property which does not entail the presence or the absence of the conjectured property.'

We can now define the concept of relative chance:

The probability of A on evidence H is called the chance, relative to ϕ_0, that the positive instance x of H will be a positive instance of A, if x has no property G, which belongs to ϕ_0, or is a presence-function of members of ϕ_0, and which is such that the probability of A on evidence $H\&G$ is different from the probability of A on evidence H alone.

The idea of chance is related to the ideas of independence (dependence) and irrelevance (relevance) which we dealt with in the preceding chapter. That the probability of A on evidence H is the chance, relative to ϕ_0, that x which is H will also be A, means that A is independent for probability in H of any property of x, which belongs to ϕ_0 or is a presence-function of members of ϕ_0. Or, speaking in less rigorous terms: it means that ϕ_0 contains no information about x which is relevant to the probability of the occurrence of A in x.

If, in the above definition of relative chance, we suppress all reference to ϕ_0, we get the following definition of absolute chance:

The probability of A on evidence H is called the chance that the positive instance x of H will be a positive instance of A, if x has no property G such that the probability of A on evidence $H\&G$ is different from the probability of A on evidence H alone.

Thus if the probability of A on evidence H is the chance that x which is H will also be A, then there is no further information beside H about x which is relevant to the probability of the occurrence of A in x. In other words: the chance is the probability that x will be A on *all relevant information* as evidence. (The name H, naturally, may denote a conjunction of several properties.)

The notion of absolute chance seems to enter into contexts of the following type:

Suppose we predict on the basis of H the occurrence of A in a given thing x. We ask: what is the probability of the prediction?

It is clear that, when raised in this way, the question of probability cannot be satisfactorily answered *merely* by pointing to the probability of A on evidence H. For if x has in addition to H a property G also, such that the probability of A on evidence $H\&G$ is different from the probability of A on evidence H alone, we should not call the probability on evidence H the probability of the prediction. (A may, *e.g.*, have a high probability on evidence H and yet it may be very improbable that our prediction about the occurrence of A in *this* instance of H will turn out true, considering that it is also in instance of G.) Further, if x has in addition to H and G a property G' also, such that the probability of A on evidence $H\&G\&G'$ is different from the probability of A on evidence $H\&G$, then we should *not* call the probability on evidence $H\&G$ the probability of the prediction either. And so on. If, on the other hand, x has in addition to H (or to H and G or to H and G and G', *etc.*) no property G (or G', *etc.*), such that the probability of A on evidence $H\&G$ ($H\&G\&G'$, *etc.*) is different from the probability of A on evidence H (or $H\&G$ or $H\&G\&G'$, *etc.*), then it is plausible to call the probability on evidence H (or $H\&G$ or $H\&G\&G'$, *etc.*) the probability of the prediction.

The probability of the prediction, in other words, is the probability of the conjectured property on *all relevant information* about the thing as evidence. It is the probability which uses as evidence a field of measurement in which the probability of the conjectured property is the chance of its occurrence in the thing.

3. Random Distribution, Equal Possibility, Rational Degree of Belief, and Chance

The distinction between a mere probability and a chance probability may be useful for the purpose of illuminating the distinctions between a mere relative frequency and a relative frequency in a *random* sequence, between a ratio of mere possibilities and a ratio of *equal* possibilities, and between a mere degree of belief and a *rational* degree of belief.

Suppose we interpret the probability of A on evidence H as meaning the proportion or relative frequency of A's among positive instances of H or, if H is (denumerably) infinite, among members of a sequence H, R.

The logical product of H with G may be said to constitute a sub-property of the main property H, and a sub-sequence of the main sequence H, R. We may also refer to G as a selection principle.

That the probability of A on evidence H&G is different from the probability of A on evidence H alone, means, on the Frequency Interpretation, that the relative frequency of A's among positive instances of the sub-property H&G or members of the sub-sequence H&G, R, is different from the relative frequency of A's among positive instances of the main property H or members of the main sequence H, R. Or else it means that G can be used for selecting a sub-property or a sub-sequence in which the proportion of positive instances of the conjectured property differs from its proportion in the main property or the main sequence.

We can now define irregular or random distribution as follows:

The distribution of A in H (or H, R) is called random relative to ϕ_0, if there is no property G, which belongs to ϕ_0, or is a presence-function of members of ϕ_0, and which is such that the relative frequency of A in $H\&G$ (or $H\&G$, R) is different from the relative frequency of A in H (or H, R).

If, in the above definition of relative randomness, we suppress all reference to ϕ_0, we get the following definition of absolute randomness:

The distribution of A in H (or H, R) is called random, if there is no property[1] G such that the relative frequency of A in $H\&G$ (or $H\&G$, R) is different from the relative frequency of A in H (or H, R).

Thus (relative and absolute) random distribution implies (relative and absolute) chance. If probability is interpreted as frequency, and if there is no property G such that the relative frequency of A in $H\&G$ (or $H\&G$, R) is different from the relative frequency of A in H (or H, R), then there cannot exist any information about any single positive instance x of H which would be relevant to the probability of A in H. The probability of A on evidence H is, in other words, for each positive instance of H the chance that it will be a positive instance of A also. It is plausible to call random distribution chance distribution.

An assertion of random distribution is, however, somewhat stronger than an assertion of chance.

Take two positive instances of H. Let us call them x and y. Let it be the case that x has no property such that the probability of A on H *and* this property as evidence is different from the probability of A on evidence H alone. Let it further be the case that y has a property such that the probability of A on H *and* this property as evidence *is* different from the probability of A on evidence H alone. Let us call this property of y G. It follows that x must be a negative instance of G. Since the probability of A on evidence $H\&G$ *is* different from the probability of A on evidence H alone, and since the probability of A on evidence $H\&\bar{G}$ is *not* different from the probability of A on evidence H alone, it follows that the probability of A on evidence $H\&G$ is different from the probability of A on evidence $H\&\bar{G}$. From the treatment of independence in Chap. VII, §12 we know that the probability of A on evidence $H\&\bar{G}$ can be the same as on evidence H alone,

---

[1] For the meaning of 'no property' cf. above p. 226.

but different from the probability on evidence $H\&G$, only in the extreme case, when the probability of $\overline{G}$ on evidence H is 1.

Thus 'normally,' *i.e.*, this extreme case being excluded, (relative and absolute) chance implies (relative and absolute) random distribution.

It was suggested above (Chap. VII, §1) that the additional demand for random distribution, which some authors associate with the frequency definition of probability, might be understood against the background of a demand or ideal of *rationality* in estimations of probability. In view of the relation between random distribution and chance, this ideal of rationality becomes identical with a demand that estimations of probability, when applied for the purpose of predicting individual occurrences of a conjectured property, should be based on *all relevant information.*

Suppose next that we interpret the probability of A on evidence H as meaning the ratio of the number of possible ways in which A and H can co-exist, to the number of possible ways in which H can exist with regard to A. (For an explanation see above p. 217 f.)

That the probability of A on evidence $H\&G$ is different from the probability of A on evidence H alone, means, on the Possibility Interpretation, the following:

The perfect disjunctive normal denotations of the two evidence-properties in terms of all atomic names in the context, contain different proportions of conjunction-names which entail the name of the conjectured property. Or, to put it otherwise: that G alters the probability, means that the original disjunction of alternative ways in which the evidence-property can exist is replaced by a new disjunction with a different balance of ways which are 'favourable' to the occurrence of the conjectured property. (Cf. above p. 217 f.)

The logical mechanism of 'disturbing the balance' needs some further elucidation.

In the perfect disjunctive normal form H appears as a disjunction of n conjunction-names—let us call them $H_1, \ldots, H_n$. Some of them entail A, some do not. That the conjunction of G with the disjunction of $H_1, \ldots, H_n$ alters the balance of favourable and unfavourable alternatives, means one of two things. Either it means that some of the alternatives become 'extinguished,' *i.e.*, contradictory or empty. *E.g.*, if G is identical with $\overline{H}_1$, then the alternative $H_1 \& G$ does not occur in the perfect disjunctive normal form of $H \& G$. Or it means that some of the alternatives become 'split up,' *i.e.*, sub-divided into a number of new alternatives. *E.g.*, if G is identical with $\overline{H}_1 v G_1 v G_2$, then the alternative H_1 becomes the alternative $H_1 \& G_1 v H_1 \& G_2$, and the alternative H_2 the alternative $H_2 \& \overline{H}_1 v H_2 \& G_1 v H_2 \& G_2$, in the perfect disjunctive normal form of $H \& G$.

That G alters the probability, thus means that the original disjunction of alternative ways in which the evidence-property can exist, is replaced by a new disjunction, in which some of the original alternatives do not occur at all or are sub-divided into new alternatives.

We can now give a meaning to possibilities being equal.

The alternative ways in which the evidence-property can exist with regard to the conjectured property are called equally possible, relative to ϕ_0, if there is no property, which belongs to ϕ_0 or is a presence-function of members of ϕ_0, and which is such that its conjunction with the evidence-property alters the proportion of ways in which the evidence-property and the conjectured property can co-exist.

If we suppress reference to ϕ_0, we get:

The alternative ways in which the evidence-property can exist with regard to the conjectured property are called equally possible, if there is no further property,[1] such that its conjunction with the evidence-property alters the proportion of ways in which the evidence-property and the conjectured property can co-exist.

Speaking roughly, and considering what was said above, the equal possibility of the alternatives means that there is no way of extinguishing or splitting up some of them so as to effect an alteration in the balance of favourable and unfavourable cases.

Thus (relative or absolute) equipossibility implies (relative or absolute) chance. It is easy to see that an assertion of equipossibility is stronger than an assertion of chance in the same

---

[1] For the meaning of 'no property' cf. above. (p. 226.)

way as that in which an assertion of randomness is stronger than an assertion of chance, but that 'normally', *i.e.*, extreme probabilities being excluded, the assertions are equally strong.

It was suggested above (Chap. VII, §1) that the additional demand for equipossibility, which is traditionally associated with the possibility definition of probability, might be understood against the background of a demand or ideal of *rationality* in estimations of probability. In view of the above relation between equal possibility and chance, this ideal of rationality becomes identical with the demand to base probability on *all relevant information.*

Let us consider what the definition of possibilities being equal, suggested above, would come to 'in practice.' Take a coin which is biased in favour of, say, 'heads.' If the probability of 'heads' in tossing with this coin is to be accounted for in terms of possibilities at all, then we must be able to replace the disjunction of the two possibilities 'heads' and 'tails' by a new disjunction of possibilities, a greater number of which are favourable to 'heads' than to 'tails.' How this is to be done, is not quite easy to see; it is sometimes suggested that the bias alters the proportion of possible ways, favourable to the respective results, for the action of mechanical forces upon the coin when twisting in the air. The plausibility of this suggestion need not concern us here. It will suffice to observe, that if the possibility definition is to have an application at all in this case, *some* such replacement of 'heads' and 'tails' by new possibilities must take place. This replacement would explain why bias creates an inequality in the two alternatives 'heads' and 'tails.'

What, however, would be the criterion for deciding whether these new alternatives are themselves equal or unequal? To this question two answers appear possible.

The first answer is that the possibilities are equal if they represent a set of 'ultimate' possibilities, none of which can be extinguished or split up so as to give a new set of possibilities with a different balance of favourable and unfavourable cases. This answer would use as its criterion our above definition of equal possibilities as possibilities which make a probability calculated on their basis a chance.

The second answer is that the possibilities are equal if they give a *true* account of the probability of 'heads' in tossing. This answer would mean that probability, or some definition of probability other than the possibility definition, is made the standard of equality in possibilities. We adjust our estimations of equal possibilities on the basis of estimations of probability, and not *vice versa*.[1]

In cases of the biased coin type, the second answer appears to be the more important in practice. The first is resorted to, if at all, merely as a hypothetical construction for the purpose of assimilating a case, in which the possibility definition of probability has no plausible application at all, to other analogous cases, in which the application of the definition is obvious and plausible.

Suppose finally that we suggested an interpretation of the probability of A on evidence H as meaning the degree of belief which we entertain in the occurrence of A on knowing that H is present.

Let us, for the sake of argument, assume that partial belief can be measured and that we believe A to degree p on sole evidence H. Let us assume that there is a positive instance x of H which also has the feature G, and that we believe A to degree q on evidence $H\&G$, p and q being different. We might then say to ourselves that it is *not* rational in this instance to believe A to degree p, since taking into the account that the instance also has the feature G would make us believe differently. Suppose, on the other hand, that there is an instance x of H which has no feature G such that we should believe A to degree q on evidence $H\&G$, p and q being different. We might then say to ourselves that it *is* rational to believe A in this instance to degree p, since no further information about the instance would make us believe differently.

I think that this is a use of the word 'rational' as an attribute of belief which we sometimes employ. The 'rational' belief is the belief based on all information which is relevant to its

---

[1] The charge against the definition of probability as a ratio of equal possibilities, that it is circular, has often been made. It has also been suggested that the frequency definition of probability is the proper standard of equipossibility. Cf. Leslie Ellis, *On the Foundations of the Theory of Probabilities* : 'When we expect two events equally, we believe they will recur equally in the long run' and 'If the events are truly equally possible, they really tend to recur equally on a series of trials.'

formation. But it is not the only, and hardly even the most
frequent, use of the word. Sometimes a *rational* belief is understood to mean a *true* belief. In the case of partial belief this
means that the rational degree of belief is one which corresponds
to the (true) probability of the conjectured event. Thus truth
as a criterion of rationality would mean that probability, or
some definition of probability other than the psychological
definition, is made the standard of rationality in our partial
beliefs.[1]

4. Chance and Determinism

The idea of chance has an important relation to the idea of
determinism (Determined Property, Principle of Determinism)
which we introduced and discussed in §2 of Chap. III.

A property is determined in ϕ_0 if, in every positive instance
of the property, there is present a Sufficient Condition of it
which belongs to ϕ_0, or is a presence-function of members of ϕ_0.
(Cf. above p. 72.)

By a Sufficient Condition of A in ϕ_0, we understood a Sufficient Condition of A which belongs to ϕ_0, or is a presence-
function of members of ϕ_0. (Cf. above p. 71.) The sum of all
Sufficient Conditions of A in ϕ_0 we called the Total Sufficient
Condition of A in ϕ_0. (Cf. above p. 71.) That A is a Determined
Property in ϕ_0 thus also means that each positive instance of A
is a positive instance of its Total Sufficient Condition in ϕ_0.
Since anything which is a positive instance of the Total Sufficient Condition of A in ϕ_0 is a positive instance of A, it follows
that, if A is determined in ϕ_0, then any negative instance of the
Total Sufficient Condition of A in ϕ_0 is a negative instance of A,
and *vice versa*.

If G is a Sufficient Condition of A, then the probability of A
on evidence $H\&G$ is 1, and if $\bar{G}$ is a Sufficient Condition of $\bar{A}$,
then the probability of A on evidence $H\&\bar{G}$ is 0.

[1] It is fairly obvious that the traditional forms of the psychological approach to
probability were intended, not to define probability in terms of belief, but rational
belief in terms of probability, though this is frequently obscured by unhappy modes
of expression. (Cf. the quotation from de Morgan on p. 170.) It has also been
suggested that the frequency definition of probability is the proper standard of
rationality in belief. Cf. Ramsey, *op. cit.*, p. 199: 'Reasonable degree of belief=
proportion of cases in which habit leads to truth.'

Let A be a Determined Property in ϕ_0. Every positive instance x of H is either a positive or a negative instance of the Total Sufficient Condition of A in ϕ_0. Let us call the Total Sufficient Condition of A in ϕ_0 G. Then the probability of A on evidence $H \& G$ is 1 and the probability of A on evidence $H \& \bar{G}$ is 0.

Thus if A is a Determined Property in ϕ_0, every positive instance x of H either has a property such that the probability of A on H *and* this property as evidence is 1, or has a property such that the probability of A on H *and* this property as evidence is 0. It follows that the probability of A on evidence H can be the chance that an x which is H will also be A, only if either the probability of A on evidence H is 1 and x is a positive instance of the Total Sufficient Condition of A in ϕ_0, or the probability of A on evidence H is 0 and x is a negative instance of the Total Sufficient Condition of A in ϕ_0.

This can also be expressed less rigorously by saying that if A is a Determined Property in ϕ_0, then, for any H, the chance that an x which is H will also be A is either 1 or 0.

Suppressing reference to ϕ_0, we got (p. 76) the following definition of absolute determinism:

A property is a Determined Property if it is determined in the set of all properties of which it is logically totally independent.

It immediately follows, that if A is a Determined Property, then, for any H, the chance that an x which is H will also be A is either 1 or 0.

By the Principle of Determinism we understood the idea that all members of a certain Universe of Properties are Determined Properties.

It follows from this that, in a Universe of Properties for which the Principle of Determinism is valid, the chance that a thing will exhibit a certain property is either 1 or 0. This might also be expressed by saying that in such a universe chance does not exist.

Similarly for random distribution. If A is a Determined Property, then it cannot in an absolute sense be said to be distributed in a random or irregular way over the members of any field of measurement H. And in a universe for which the Principle of Determinism is valid, there is no room for random distribution at all.

These facts, however, do not exclude a Determined Property from possessing a (not-extreme) chance for its occurrence in a given positive instance of another property, or from being randomly distributed over the positive instances of another property, *relative to some set of properties ϕ_0*.

Thus absolute determinism does not exclude relative chance (or randomness or equipossibility). The existence of (relative) chance under the rule of (absolute) determinism has a peculiar epistemological significance in cases where the range ϕ_0, relative to which there is a chance, answers to a certain state of actual knowledge.

Note.—The importance of random distribution to the notion of probability was stressed by von Mises. His definition of probability as a (limiting) frequency in a random sequence could also be described as the frequency definition to which has been added the demand to base estimations of probability on *all relevant information*.

The subsequent discussion of random distribution has mainly centred round the problem, how to give a consistent formulation to the intended content of von Mises' notion of a *Kollektiv*. This discussion, it seems to me, has suffered from two limitations. First, it has tackled the problem involved in the context exclusively from the point of view of the Frequency Interpretation of probability. Secondly, it has generally been concerned only with that kind of probability which we have called intensional, *i.e.*, probability-propositions which on the Frequency Interpretation are decidable on formal grounds.

It appears, however, that the problem of random distribution is a special case of the more general problem of chance or probability based on all relevant information, and that the classical problem of equipossibility is another special case of it. It also seems to me that the problem must be separately treated for what we have here called extensional probability, and that it then has an important connexion with the idea of determinism.

A treatment of the idea of random distribution on lines similar to those outlined here was attempted by me in the paper *On Probability* in Mind *49* (1940) and further developed in my publication *Ueber Wahrscheinlichkeit* (1945).

Chapter Nine

PROBABILITY AND LAWS OF NATURE

1. The Concept of Real Inductive Probability

IN the present chapter we shall deal with probability as a genuine attribute of laws. For this type of probability we have reserved the name (Real) Inductive Probability.

The idea of (Real) Inductive Probability has been the object of much controversy. Views as to its nature may be schematically divided into three groups:

i. The nihilistic views deny altogether the possibility of attributing probability to inductive conclusions of the second order, at least in so far as they have the numerically unrestricted character of laws. Of such views one might distinguish two variants, according to the way in which the word 'possibility' is understood, *viz.*

α. views rejecting the concept of Inductive Probability for philosophical or logical reasons, *e.g.*, because it is thought to involve contradictions, and

β. views regarding the concept as being useless in practice.

ii. The dualistic views take the idea of Inductive Probability to be philosophically and logically legitimate, though somehow different from what may vaguely be called 'ordinary' probability. One might distinguish between two main variants of this opinion also, according to what sort of difference between inductive and 'ordinary' probability is contemplated, *viz.*

a. views assuming that the concept of Inductive Probability is not subject to the rules of our Calculus of Probability (though perhaps to those of some other logico-mathematical formalism), and

β. views assuming that the concept of Inductive Probability has the logico-mathematical structure of the calculus in common with 'ordinary' probability and that the difference is in the *interpretation* of the calculus. It might, *e.g.*, be suggested that 'ordinary' probability is a frequency, and Inductive Probability a ratio of possibilities.

iii. The monistic views assume not only that the idea of Inductive Probability is legitimate, but also that its difference, if any, from 'ordinary' probability is not a difference in the interpretation of a common axiomatic frame.

We shall attempt to show that views of type ia, or nihilistic views of the stronger type, are not justified. This means that the concept of Inductive Probability need not be rejected as contrary to logic.

In dealing with the concept of Inductive Probability we shall disregard views of type ii$a$, or dualistic views of the stronger type. This means that the concept of Inductive Probability is here treated within the framework of our Calculus of Probability. It does not mean, however, that a different treatment might not be possible and for some purposes even useful.

Further, we shall attempt to show that the difference between 'ordinary' and Inductive Probability can be accounted for without resort to different interpretations of the calculus. This means that we shall disregard also views of type iiβ, or dualistic views of the weaker type and accept a view of the monistic type. It does not mean, however, that another and more radical account of the difference might not be possible and for some purposes even necessary.

There still remains to be considered the relation of iii to iβ. As will be seen, there are strong reasons for combining our monistic view of 'ordinary' and Inductive Probability with a nihilistic attitude as regards the practical importance of Inductive Probability. These reasons, moreover, are closely connected with the way in which we here try to account for the difference between the two types of probability.

The discussion of the nature of (Real) Inductive Probability will be continued in the next section. The remaining portion of this chapter will be devoted to what we propose to call a reconstructive examination of certain familiar arguments of (Real) Inductive Probability.

2. *The Concept of Real Inductive Probability (Continued)*

Probability, we said (p. 172), is an attribute of propositions.

In order to make clear that the views under iα can be rejected, and that those of iiα can be evaded or disregarded, it will be our first duty to show in what way the probability-relation P of our calculus is applicable to Universal Implications and Equivalences.

In dealing with probability, we also said (p. 172) that a certain analysis of propositions will be presupposed. By the probability of something we understand the probability that a thing has a certain property.

Probability as an attribute of laws, therefore, presupposes here what we have called (p. 37) an Aristotelian analysis of Universal Implications and Equivalences. According to the view adopted in this inquiry (p. 37), every proposition is capable of this sort of analysis. In the case of Universal Implications and Equivalences, their Aristotelian analysis is as follows:

Let ϕ_H denote the (second-order) property of being a Necessary Condition of the property denoted by H. Let ϕ_A denote the property of being a Sufficient Condition of the property denoted by A. The Universal Implication $H \subset A$ can be regarded either as predicating the second-order property ϕ_H of the first-order property A or as attributing the second-order property ϕ_A to the first-order property H. (These are two principal ways of effecting an Aristotelian analysis of the Universal Implication $H \subset A$; there are also other ways, but they need not be discussed here.)

Let ψ_H denote the property of being a Necessary-and-Sufficient Condition of the property denoted by H. Let ψ_A

denote the property of being a Necessary-and-Sufficient Condition of the property denoted by A. The Universal Equivalence $H \equiv A$ can be regarded either as attributing the second-order property ψ_H to the first-order property A or as attributing the second-order property ψ_A to the first-order property H. (These are two principal ways of effecting an Aristotelian analysis of the Universal Equivalence $H \equiv A$; there are also other ways, but they need not be discussed here.)

Probability, we said (p. 172), is relative to evidence. We took the view that the evidence is a property of *the same* thing as the property, the presence of which is being conjectured.

Let ϕ_0 denote membership of a set of properties. ϕ_0 is thus a property of a property. We take ϕ_0 to belong to the same universe as the four properties ϕ_H, ϕ_A, ψ_H and ψ_A defined above. It will not be ussumed that the members of ϕ_0 are logically totally independent.

Then, according to what was said in Chap. VII, §1, any one of the four properties ϕ_H, ϕ_A, ψ_H and ψ_A in combination with ϕ_0 and any real number p will constitute either a positive or a negative order of the probability-relation P.

The probability-expression $P(\phi_H, \phi_0, p)$ means: p is the probability that a property will be a Necessary Condition of H on the evidence that it belongs to ϕ_0. For instance: p is the probability of the law $H \subset A$ on the evidence that A belongs to ϕ_0.

The probability-expression $P(\phi_A, \phi_0, p)$ means: p is the probability that a property will be a Sufficient Condition of A on the evidence that it belongs to ϕ_0. For instance: p is the probability of the law $H \subset A$ on the evidence that H belongs to ϕ_0.

The probability-expression $P(\psi_H, \phi_0, p)$ means: p is the probability that a property will be a Necessary-and-Sufficient Condition of H on the evidence that it belongs to ϕ_0. For instance: p is the probability of the law $H \equiv A$ on the evidence that A belongs to ϕ_0.

The probability-expression $P(\psi_A, \phi_0, p)$ means: p is the probability that a property will be a Necessary-and-Sufficient Condition of A on the evidence that it belongs to ϕ_0. For instance: p is the probability of the law $H \equiv A$ on the evidence that H belongs to ϕ_0.

By saying that nihilistic views of Inductive Probability of the stronger type can be rejected, and that the stronger type of dualistic views can be evaded, we shall here mean simply that there is an Aristotelian analysis of Universal Implications and Equivalences which makes it possible to define the four properties and arrive at the four probability-expressions above.

It seems to be generally true of propositions that there are different ways of viewing them in terms of predicate and subject. Sometimes, however, there is one way which immediately presents itself as the 'natural' one. Sometimes again there are two or more ways which appear in themselves equally 'natural.' Probability as an attribute of propositions seems ordinarily to be an attribute of propositions of the first kind. Universal Implications and Equivalences, however, are decidedly of the second kind. The proposition, *e.g.*, that the next throw with a certain die will be a one 'naturally' attributes the property of being a one to the next throw. But the proposition, *e.g.*, that all ravens are black does nòt 'naturally' attribute to the property of being a raven the second-order property of being a Sufficient Condition of blackness; nor does it 'naturally' attribute to the property of being black the second-order property of being a Necessary Condition of ravenness.

The concept of Inductive Probability, as dealt with in this inquiry, is thus ambiguous in the sense that it must always be explicitly associated with one of an alternative number of analyses of the law to which it is being applied. This feature marks a first conspicuous difference between Inductive Probability and probability which, in a vague sense, may be called 'ordinary.'

The evidence or field of measurement of an Inductive Probability can be conveniently described as a set of possible conditioning properties of a given conditioned property.

Ordinarily, in attributing probability to something, there is at least some vague indication as to the evidence or field of measurement, relative to which probability is contemplated. (Though it might happen that part of the contemplated evidence is suppressed in loose modes of speech.) *E.g.*, in speaking about the probability of getting a one in the next throw with a certain die it is understood that the probability is (at least) relative to the property of being a throw with this die as evidence.

In attributing probability to laws, however, there seems normally to be no indication whatsoever as to the evidence or field of measurement relative to which the probability is contemplated. To say, *e.g.*, that the probability of a law is relative to all relevant information or to 'our present state of knowledge' as evidence does not alter the situation unless we are told, vaguely at least, *what* this information or state of knowledge is.

The normally unspecified character of the evidence or the fields of measurement is a second feature of Inductive Probability which distinguishes it from probability which may, in a vague sense, be called 'ordinary.' This feature, moreover, is at least partly responsible for the fact that in the realm of Inductive Probability precise numerical evaluations are not normally regarded as possible.

Further problems concerning the choice and possible specification of the fields of measurement of Inductive Probability will not be discussed in this inquiry.

$$* \qquad * \qquad * \qquad * \qquad *$$

In the Frequency Interpretation $P(\phi_H, \phi_0, p)$ means: p is the proportion of Necessary Conditions of H among members of ϕ_0. Similarly, we interpret $P(\phi_A, \phi_0, p)$ and $P(\psi_H, \phi_0, p)$ and $P(\psi_A, \phi_0, p)$ as stating a proportion of actual conditions among initially possible conditions.

On the Possibility Interpretation $P(\phi_H, \phi_0, p)$ means: p is the ratio of possible ways in which a property can be both a Necessary Condition of H and a member of ϕ_0, to the number of possible ways in which it can be a member of ϕ_0. (Cf. above p. 218.) We interpret $P(\phi_A, \phi_0, p)$ and $P(\psi_H, \phi_0, p)$ and $P(\psi_A, \phi_0, p)$ similarly. The interpretation is of no particular interest, unless ϕ_0 is viewed as a presence-function of other properties.

On the Psychological Interpretation $P(\phi_H, \phi_0, p)$ would mean: p is the degree to which we believe that a property is a Necessary Condition of H on the evidence that it is a member of ϕ_0. We interpret $P(\phi_A, \phi_0, p)$ and $P(\psi_H, \phi_0, p)$ and $P(\psi_A, \phi_0, p)$ likewise.

Against the Frequency Interpretation of Inductive Probability the following objection has been made:

Universal Implications and Equivalences, in the numerically unrestricted sense of Laws of Nature, have no verifying instances. Consequently, it is not possible to know for certain, of any property which belongs to a set of possible conditions of a given property, that it belongs to the set of actual conditions also. (Though we may know that it does *not* belong to it.) Or as Popper[1] has expressed it: 'Dieser Versuch scheitert . . . daran, dass wir von einer Wahrheitshäufigkeit innerhalb einer Hypothesenfolge schon deshalb nicht sprechen können, weil wir ja Hypothesen zugestandenermassen nicht als "wahr" kennzeichnen können. Denn könnten wir das—wozu brauchen wir dann noch den Begriff der Hypothesenwahrscheinlichkeit?'

The fact, however, that Universal Implications and Equivalences are not verifiable on the basis of their instances is no objection against the legitimacy (logical consistency) of the idea that, of a set of such propositions, a certain proportion are true propositions. Or to put it in a different way: it is no objection against the legitimacy of the idea that of a set of possible conditions of a property a certain proportion are actual conditions of the property. And that is all that is contained in a proposition of Inductive Probability on the Frequency Interpretation.

The objection, therefore, has no force against the 'possibility' of the Frequency Interpretation. This will justify us in disregarding also the views under iiβ of the preceding section or the weaker type of what we called dualistic views of Inductive Probability.

The real effect of the objection is, it seems, not to invalidate the Frequency Interpretation of Inductive Probability, but to point out a very important difference—*within* the interpretation in question—between Inductive Probability and probability which, in a vague sense, may be called 'ordinary'. This difference is in the strength of the basis of statistical observations in support of a probability proposition. Ordinarily, *e.g.*, in the case of games of chance or of insurance against risks, these observations will be verified propositions concerning relative frequencies among a finite number of things. In the case of Inductive Probability, however, we have to content ourselves

[1] *Logik der Forschung* (1935), p. 192.

with mere assumptions as to relative frequencies even in finite sections of the respective fields of measurement. (The only exception is the extreme case when we have succeeded in falsifying all of a finite number of possible laws.)

Thus, even in fields of measurement with a known finite number of members, the Frequency Interpretation is not what we have previously called an analysis (as opposed to inter-pretation) of the concept of Inductive Probability. This is a third difference between Inductive Probability and 'ordinary' probability.

The fact that there is no analysis of Inductive Probability in terms of frequency might also be held to be in part responsible for the other fact, already mentioned (p. 172 f.), that in the realm of Inductive Probability exact numerical evaluations are not normally regarded as possible. Estimates of Inductive Probability are notorious for their vagueness and subjectivity. A certain consensus seems to exist as to what should be con-sidered greater and less, considerable, mediocre, or minimal even in the realm of Inductive Probability. But no claims as to exact numerical evaluations appear to have much hope of be-coming universally accepted. That such should be the case is very natural indeed, considering the peculiar weakness of the support which, for reasons of logic, statistical observation can provide for Inductive Probability.

Rightly understood, therefore, the three peculiarities of Inductive Probability: the ambiguity in the analysis of the laws, the normally unspecified character of the evidence, and the impossibility of an analysis (as opposed to interpretation) in terms of frequency, might be taken as a strong indication in favour of combining the monistic view of Inductive Probability which we have decided to accept with a nihilistic view of the weaker type iβ. This nihilism, it should be observed, is justified, not so much by the lack *as such* of exact numerical estimates of probability in the case of laws, as by the fact that this lack of numerical precision is founded on the logical nature of the con-cept of Inductive Probability itself, according to our account of it. When, as is sometimes done, the vague and unsafe nature of Inductive Probability is attributed to the present imperfect state of human knowledge this is at most a partial explanation of a striking fact, the deeper reasons for which it is for logical analysis to reveal.

3. *The Argument from Confirmation*

It is one of the traditional ideas of the Logic of Inductive Probability that confirmation of a law through its instances contributes to its probability.

The probability of a law is assumed to

i. increase with the number of confirming instances,

ii. increase the more, the less probable its confirmation is in a given instance, and

iii. approach maximum probability if confirmation goes on indefinitely.

We call this the Argument from Confirmation. (Its third clause is sometimes disputed.)

It will be the purpose of the present section to examine the argument within the framework of our Calculus of Probability.

Let the law be $H \subset A$.

As will be remembered from the preceding section, probability as an attribute of a law is relative to the choice of one of a number of alternative ways of conceiving the law as a proposition attributing a predicate to a subject. The predicate may be the property of being a Necessary Condition of H and the subject the property A. The predicate may also be the property of being a Sufficient Condition of A and the subject the property H. The first predicate we denoted ϕ_H and the second ϕ_A. It will be shown that on either alternative we get the same theorem of Inductive Probability. For this reason we shall here suppress the index of ϕ.

Let ϕ, therefore, denote the set of (actual) Necessary (Sufficient) Conditions of H (A).

Supposing the instances of the conditioned property to be denumerable, we can order them into a sequence $x_1, \ldots, x_n, \ldots$. Since instances may be either positive or negative, $x_1, \ldots, x_n, \ldots$ is the sequence of all things in the universe in question. According to whether a thing is a positive or negative instance of the implication property $H{\rightarrow}A$, it affords a confirming or disconfirming instance respectively of the law $H \subset A$.

Let ϕ_0 denote a set of (initially) possible Necessary (Sufficient) Conditions of $H\,(A)$.

Let ϕ_1 denote the set of properties of which it is true that they are not absent (present) in the thing x_1 in the presence (absence) of $H\,(A)$. ϕ_1, in other words, denotes the set of remaining possible Necessary (Sufficient) Conditions of $H\,(A)$ relative to x_1. (Cf. Chap. IV, §2.)

Let ϕ_2 denote the set of properties of which it is true that they are not absent (present) in the things x_1 *and* x_2 in the presence (absence) of $H\,(A)$. ϕ_2, in other words, denotes the set of remaining possible Necessary (Sufficient) Conditions of $H\,(A)$ relative to x_1 *and* x_2.

Similarly, we introduce ϕ_3, *etc.*

In the sequence of properties $\phi_1, \ldots, \phi_n, \ldots$ every property is included in its successor. The logical product of *all* these properties is ϕ. Thus we have $lim(\phi_n, \phi)$. The set of actual conditioning properties is the limiting extension of the sets of the remaining possible conditioning properties. This is already familiar to us. (Cf. Chap. IV, §2.)

The following suppositions are introduced:

S1. The property $A\,(H)$ belongs to the set ϕ_0. Thus $E\,\phi_0$.

S2. There is a probability p that a random property, *e.g.*, $A\,(H)$, will be a Necessary (Sufficient) Condition of $H\,(A)$ on the evidence that it belongs to ϕ_0. In symbols: $P(\phi, \phi_0, p)$.

We shall call p the probability *a priori* of the law.

S3. The probability *a priori* of the law is not minimal. In symbols: $p>0$.

S4. There is a probability p_1 that a random property, *e.g.*, $A\,(H)$, will be a remaining possible Necessary (Sufficient) Condition of $H\,(A)$ relative to x_1 on the evidence that it belongs to ϕ_0. In symbols: $P(\phi_1, \phi_0, p_1)$. There is further a probability p_{n+1} that a random property, *e.g.*, $A\,(H)$, will be a remaining possible Necessary (Sufficient) Condition of $H\,(A)$ relative to x_1 and . . . and x_{n+1}, on the evidence that it belongs to ϕ_0 and is a remaining possible Necessary (Sufficient) Condition of $H\,(A)$ relative to x_1 and . . . and x_n. In symbols:

$(n)P(\phi_{n+1}, \phi_0 \& \phi_n, p_{n+1})$.

That there is a probability p_{n+1} that A (H) will be a remaining possible Necessary (Sufficient) Condition of H (A) relative to x_1 and . . . and x_{n+1} on the evidence that A (H) belongs to ϕ_0 and is a remaining possible Necessary (Sufficient) Condition of H (A) relative to x_1 and . . . and x_n, means that there is a probability p_{n+1} that x_{n+1} will afford a confirmation of the law $H \subset A$ on the evidence that A (H) belongs to ϕ_0 and that x_1 and . . . and x_n afford confirmations of $H \subset A$. We shall therefore call p_{n+1} the probability that the law will be confirmed *also* in its $n+1$:th instance.

On these four suppositions as data we can deduce a probability *a posteriori* of the law, *i.e.*, a probability that A (H) is a Necessary (Sufficient) Condition of H (A) on the evidence that A (H) belongs to ϕ_0 *and* still is a possible Necessary (Sufficient) Condition of H (A) after n confirmations of the law $H \subset A$.

$\bar{E}$ $\phi \& \phi_0$ entails $\phi_0 \subset \bar{\phi}$.

The Axiom of Inclusion and the Addition Axiom and E ϕ_0 and $\phi_0 \subset \bar{\phi}$ entail $P(\phi, \phi_0, 0)$.

The Axiom of Uniqueness and E ϕ_0 and $P(\phi, \phi_0, p)$ and $P(\phi, \phi_0, 0)$ entail $p=0$.

$p=0$ contradicts $S3$.

Thus we have proved that E $\phi \& \phi_0$.

E $\phi \& \phi_0$ entails $(n)(E$ $\phi \& \phi_n)$.

ϕ_n is identical with $\Pi \phi_n$.

Therefore, the Multiplication Principle and $(n)(E$ $\phi_0 \& \phi_n)$ and $P(\phi_1, \phi_0, p_1)$ and $(n)P(\phi_{n+1}, \phi_0 \& \phi_n, p_{n+1})$ entail (n) $P(\phi_n, \phi_0, \Pi p_n)$.

We shall call Πp_n the probability that the law will be confirmed n times in succession.

ϕ_n is also identical with $\phi v \bar{\phi} \& \phi_n$.

$(n)(\bar{E}$ $\phi \& \bar{\phi} \& \phi_n)$ is tautologous.

The Addition Principle and E ϕ_0 and $(n)(\bar{E}$ $\phi \& \bar{\phi} \& \phi_n)$ and $(n)P(\phi v \bar{\phi} \& \phi_n, \phi_0, \Pi p_n)$ and $P(\phi, \phi_0, p)$ entail $(n)P(\bar{\phi} \& \phi_n, \phi_0, (\Pi p_n)\text{-}p)$.

We shall call $(\Pi p_n)\text{-}p$ the probability that the law is false but nevertheless confirmed n times in succession.

ϕ is identical with $\phi \& \phi_n$.

The Axiom of Minimum Probability and E ϕ_0 and $(n)P(\bar{\phi} \& \phi_n, \phi_0, (\Pi p_n)\text{-}p)$ entail $(n)((\Pi p_n)\text{-}p \geqslant 0)$.

$(n)((\Pi p_n)\text{-}p \geqslant 0)$ and $p>0$ entail $(n)(\Pi p_n>0)$.

The Multiplication Principle and $(n)(E\ \phi_0\&\phi_n)$ and $(n)P(\phi\&\phi_n,\phi_0,p)$ and $(n)P(\phi_n,\phi_0,\Pi p_n)$ and $(n)(\Pi p_n>0)$ entail $(n)P(\phi,\ \phi_0\&\phi_n,\ p:\Pi p_n)$.

$p:\Pi p_n$ is the probability *a posteriori* of the law.

Let us compare $p:\Pi p_{n+1}$ and $p:\Pi p_n$. If $p>0$ and $p_{n+1}<1$, then $p:\Pi p_{n+1}>p:\Pi p_n$.

Let us consider the difference $p:\Pi p_{n+1}-p:\Pi p_n$. It is inversely proportional to p_{n+1}.

The Axiom of Continuity and $E\ \phi_0$ and $P(\phi,\ \phi_0,\ p)$ and $(n)P(\phi_n,\ \phi_0,\ \Pi p_n)$ and $lim(\phi_n,\ \phi)$ entail $lim(\Pi p_n,\ p)$.

$p>0$ and $lim(\Pi p_n,\ p)$ entail $lim(p:\Pi p_n,\ 1)$.

The probability *a posteriori* thus approaches 1 as a limit.

$lim(\Pi p_n,\ p)$ is identical with $lim((\Pi p_n)\text{-}p,\ 0)$.

That the probability *a posteriori* approaches 1 as a limit thus means the same as

α. that the probability that the law will be confirmed n times in succession approaches as a limit the probability *a priori* of the law, and

β. that the probability that the law is false but nevertheless confirmed n times in succession approaches 0 as a limit.

It should be observed that the convergence of the probability *a posteriori* towards maximum probability depends upon the convergence of the sets of remaining possible conditioning properties towards the set of actual conditioning properties. This latter convergence requires that the sequence of things $x_1,\ \ldots,\ x_n,\ \ldots$, on the basis of which the sequence of properties $\phi_1,\ \ldots,\ \phi_n,\ \ldots$ is defined, is the sequence of *all* things in the universe in question. In practice, however, the things affording confirmations of a law are likely to be restricted to some part of the universe only. In so far as this is the case, convergence towards maximum probability cannot be proved.

Our examination of the Argument from Confirmation has so far taken us to the following Theorem of Confirmation:

Let there be a not empty set ϕ_0 of initially possible Necessary (Sufficient) Conditions of the property $H\ (A)$. Suppose that there is a not minimal probability *a priori* of the law that $A\ (H)$ is an actual Necessary (Sufficient) Condition of $H\ (A)$ on the evidence that $A\ (H)$ belongs to ϕ_0. Suppose further that there is a probability that the law will be confirmed in the first instance, in the second instance *also*, *etc.* Then the probability *a posteriori* of the law that A (H) is an actual Necessary (Sufficient) Condition of H (A) on the evidence that A (H) belongs to ϕ_0 and *still is* a possible Necessary (Sufficient) Condition of H (A) after n confirmations of the law,

i. increases with each instance affording a confirmation, provided it is not maximally probable that the law will be confirmed in this instance *also*;

ii. increases the more with each instance affording a confirmation, the less probable it is that the law will be confirmed in this instance *also*, and

iii. approaches maximum probability, if confirmation goes on indefinitely.

There is a corresponding theorem for the law $H \equiv A$.

4. The Argument from Confirmation (Continued)

If the results of the preceding section are viewed in the light of the Frequency Interpretation of our Calculus of Probability, then the logical mechanism responsible for the increase in Inductive Probability through confirmation will be found to be identical with the logical mechanism of Induction by Elimination.

As will be remembered, ϕ_0 is a set of initially possible Necessary (Sufficient) Conditions of the property H (A). $\phi_1, \ldots, \phi_n, \ldots$ are the sets of remaining possible Necessary (Sufficient) Conditions of the property H (A). ϕ is the set of actual Necessary (Sufficient) Conditions of the property H (A).

The probability *a priori* of the law $H \subset A$ means, on the Frequency Interpretation, the proportion of actual Necessary (Sufficient) Conditions of H (A) among the initially possible conditions. How ϕ_0 should be chosen so that this proportion is not minimal is an interesting problem of methodology which, however, we shall not discuss here.

The probability that the law $H \subset A$ will be confirmed in the first instance means, the proportion of remaining possible

Necessary (Sufficient) Conditions of H (A) relative to x_1, among the initially possible conditions.

The probability that the law $H \subset A$ will be confirmed *also* in the $n+1$:th instance means, the proportion of remaining possible Necessary (Sufficient) Conditions of H (A) relative to the first $n+1$ things, among the remaining initially possible conditions relative to the first n things.

The probability that the law $H \subset A$ will be confirmed n times in succession means, the proportion of remaining possible Necessary (Sufficient) Conditions of H (A) relative to the first n things, among the initially possible conditions.

This proportion is the product of the proportion of remaining possible Necessary (Sufficient) Conditions of H (A) relative to the first thing, among the initially possible conditions and . . . and the proportion of remaining possible Necessary (Sufficient) Conditions of H (A) relative to the first n-1 things, among the remaining initially possible conditions relative to the first n things.

The probability *a posteriori* of the law $H \subset A$ means the proportion of actual Necessary (Sufficient) Conditions of H (A) among the remaining initially possible conditions relative to the first n things.

This last proportion is the ratio of the proportion of actual Necessary (Sufficient) Conditions of H (A) among the initially possible conditions, to the proportion of the remaining possible Necessary (Sufficient) Conditions of H (A) relative to the first n things among the initially possible conditions.

The proportion of actual conditions among the initially possible conditions is a constant. The alterations in the probability *a posteriori* are thus a function of alterations in the proportion of remaining possible conditions relative to the first n things among the initially possible conditions.

The proportion of remaining possible conditions relative to the first $n+1$ things among the initially possible conditions is equal to, or smaller than, the proportion of remaining possible conditions relative to the first n things among the initially possible conditions.

If the proportions are equal, then the proportion which answers to the *a posteriori* probability is not altered when we pass from the first n to the first $n+1$ things.

If the proportions are not equal, then the proportion which answers to the *a posteriori* probability is increased when we pass from the first n to the first $n+1$ things.

That the proportions are not equal means the same as that the proportion of remaining possible conditions relative to the first $n+1$ things among the remaining initially possible conditions relative to the first n things, is not 1.

Two cases must now be distinguished: either ϕ_0 is finite or ϕ_0 is denumerably infinite.

If ϕ_0 is finite, then to say that the last-mentioned proportion is not 1, means that there is at least one initially possible condition which is a remaining possible condition relative to the first n things but *not* relative to the first $n+1$ things. Thus to say that the proportion is not 1 means that the $n+1$:th thing excludes or eliminates at least one of the initially possible Necessary (Sufficient) Conditions of $H(A)$ from being an actual Necessary (Sufficient) Condition of $H(A)$.

If ϕ_0 is denumerably infinite, then to say that the proportion is not 1, means that a perceptible proportion (p. 80) of the remaining possible conditions relative to the first n things are *not* remaining possible conditions relative to the first $n+1$ things. Thus to say that the proportion is not 1 entails, but is not identical with saying that the $n+1$:th thing excludes or eliminates at least one of the initially possible Necessary (Sufficient) Conditions of $H(A)$ from being an actual Necessary (Sufficient) Condition of $H(A)$.

We shall henceforth always understand 'elimination' in the stronger sense of elimination of a *perceptible proportion* (and not merely *at least one*) of the remaining possible conditions.

Thus, on the Frequency Interpretation, increase in the probability *a posteriori* of a law means elimination.

The difference between the proportions which answer to the *a posteriori* probabilities for $n+1$ and n respectively is inversely proportional to the proportion of remaining possible conditions relative to the first $n+1$ things, among the remaining initially possible conditions relative to the first n things.

If ϕ_0 is finite, then the last-mentioned proportion is the smaller, the greater the number of initially possible Necessary (Sufficient) Conditions of $H(A)$ that the $n+1$:th thing excludes or eliminates from being actual Necessary (Sufficient) Conditions of $H(A)$.

If ϕ_0 is denumerably infinite, then the proportion is the smaller, the greater is the proportion of remaining possible Necessary (Sufficient) Conditions of $H(A)$ relative to the first n things, which are *not* remaining possible Necessary (Sufficient) Conditions of $H(A)$ relative to the first $n+1$ things.

Thus, on the Frequency Interpretation, increase in the probability *a posteriori* of a law is proportionate to the eliminative efficiency of the confirming instances.

It is a well-known fact that common sense is not prepared to regard all confirming instances of a law as equally effective in raising its probability. It is also the case that, as a rule, we regard the first few successful confrontations of a law with experience as more important from the point of view of probability than even a large number of later confirmations.

It appears to me very plausible to hold that this unequal power of raising the probability of a law should be attributed to inequality in the eliminative efficiency of the confirming instances. It is also reasonable to ascribe the relative importance of the initial confirmations to the fact that, as a rule, the eliminative efficiency decreases as more and more confirmations are made. It becomes increasingly difficult 'to vary the circumstances.'

It is hard to see what could be meant by the 'improbability' of the confirmation of a law in a new instance except that the new instance invalidates a great number of laws, and that therefore the law in question runs great 'risks' of becoming invalidated.

That the probability *a posteriori* of the law $H \subset A$ approaches maximum probability as a limit means, on the Frequency Interpretation, that the proportion of remaining possible Necessary (Sufficient) Conditions of $H(A)$ relative to the first n things among the initially possible conditions approaches as a limit the proportion of actual Necessary (Sufficient) Conditions of $H(A)$ among the initially possible conditions. It should again be noted that this is the case only if the sequence of things affording confirmations of the law successively exhausts the entire Universe of Things.

Thus, on the Frequency Interpretation, the Theorem of Confirmation runs:

Let there be a not empty set ϕ_0 of initially possible Necessary (Sufficient) Conditions of the property H (A). Suppose that a perceptible proportion of members of this set are actual Necessary (Sufficient) Conditions of H (A). (Suppose further that a proportion of members of this set remain possible Necessary (Sufficient) Conditions of H (A) relative to the first thing and, generally, that a proportion of remaining initially possible Necessary (Sufficient) Conditions of H (A) relative to the first n things remain possible Necessary (Sufficient) Conditions of H (A) relative to the first $n+1$ things.) Then the proportion of actual Necessary (Sufficient) Conditions of H (A) among the remaining initially possible conditions relative to the first n things,

i. increases with each thing, provided the thing contributes to the elimination of initially possible Necessary (Sufficient) Conditions of H (A),

ii. increases the more with each thing, the more the thing contributes to the elimination of initially possible Necessary (Sufficient) Conditions of H (A), and

iii. approaches 1 as a limit.

We shall speak of the considerations which have taken us to the Theorem of Confirmation and its Frequency Interpretation as a 'reconstructive examination' of the Argument from Confirmation.

*　　　*　　　*　　　*　　　*

The Argument from Confirmation is sometimes understood as a justification of the idea that Induction by Enumeration has an independent value apart from Induction by Elimination. On our reconstructive examination of the argument, this view becomes unwarranted. To speak of the increasing probability of a law turns out to be only a disguised manner of expressing facts about the elimination of 'concurrent' possibilities. To put it in the shortest way: the increasing probability of a law is merely another expression of the trivial fact that, as the number of possible alternative laws decreases through elimination, the proportion of true laws increases.

One might object to our reconstructive undertaking that it makes Inductive Probability utterly trivial and void of practical interest. This, I think, is true, but hardly constitutes an objection. Was it ever reasonable to expect that the Argument from Confirmation and other uses of Inductive Probability, once they had become properly clarified, would amount to anything very important and powerful from the point of view of scientific practice? It does not seem so to me. The charge against a Logic of Induction of making the concept of Inductive Probability trivial, resembles the traditional charge against the logical undertakings of Bacon and Mill of not having given much help to the actual progress of science and the discovery of new truths. The reconstructive examination has served its purpose if it has succeeded in making clearer what people actually do or *might* do when they employ arguments of Inductive Probability. If this clarification makes the arguments turn out less profound than they appeared previously, the service done by logic to the mental hygiene, so to speak, of the scientist is all the greater.

The notion of the degree of probability, or of confirmation, of a law can be defined and treated in various ways. It is by no means claimed that the lines of treatment adopted here are applicable to all cases in which this notoriously obscure notion is used. Our treatment has an immediate application to the well-known type of situation in experimental science, where there is a question of finding 'causes,' *i.e.*, conditions, of given phenomena. It has, however, no immediate application to other types of situation, *e.g.*, to estimating the probability of the Theory of Relativity. It seems to me doubtful if any systematic clarification of our notion of (Real) Inductive Probability can be given, outside the study of Induction by Elimination.

5. *The Paradox of Confirmation*

As was observed already in Chap. III, §1, any thing which is known to be a negative instance of H or a positive instance of A is *ipso facto* known to afford a confirmation of the law $H \subset A$. We referred to this fact as the Paradox of Confirmation.

PROBABILITY AND LAWS OF NATURE

We have seen that if certain conditions are fulfilled, confirmation of a law contributes to its probability. It might be asked whether this contributive effect also attaches to 'paradoxical' confirmations of $H \subset A$ via negative instances of H and positive instances of A. An affirmative answer would imply, *e.g.*, that the verification of a proposition to the effect that this is a swan might increase the probability of the law that all ravens are black. This is plainly an absurd consequence. If it were really warranted within the frame of the reconstructive examination of the Argument from Confirmation which we have pursued in the two preceding paragraphs, this would obviously be a serious blow against the success of the reconstructive undertaking itself.

It is, however, possible to show that the Paradox of Confirmation is harmless.

Let us assume that x_{n+1} is either a negative instance of H or a positive instance of A.

ϕ_{n+1} means the set of properties of which it is true that they are not absent (present) in the things x_1 and ... and x_{n+1} in the presence (absence) of H (A).

If x_{n+1} is a negative instance of H, *i.e.*, if H is absent in x_{n+1}, then every property which is not absent in x_1 and ... and x_n in the presence of H, is also not absent in x_1 and ... and x_{n+1} in the presence of H. (For, if H is absent in x_{n+1}, then every property is not absent in the presence of H in *this* instance.)

Similarly, if x_{n+1} is a positive instance of A, *i.e.*, if A is present in x_{n+1}, then every property which is not present in x_1 and ... and x_n in the absence of A, is also not present in x_1 and ... and x_{n+1} in the absence of A. (For, if A is present in x_{n+1}, then every property is not present in the absence of A in *this* instance.)

Thus, if x_{n+1} is either a negative instance of H or a positive instance of A, then the properties ϕ_{n+1} and ϕ_n are co-extensive.

$\phi_{n+1} \equiv \phi_n$ entails $\phi_n \subset \phi_{n+1}$.

$\phi_n \subset \phi_{n+1}$ entails $\phi_0 \& \phi_n \subset \phi_{n+1}$.

The Axiom of Inclusion and E $\phi_0 \& \phi_n$ and $\phi_0 \& \phi_n \subset \phi_{n+1}$ entail $P(\phi_{n+1}, \phi_0 \& \phi_n, 1)$.

The Axiom of Uniqueness and E $\phi_0 \& \phi_n$ and $P(\phi_{n+1}, \phi_0 \& \phi_n, p_{n+1})$ and $P(\phi_{n+1}, \phi_0 \& \phi_n, 1)$ entail $p_{n+1} = 1$.

$p>0$ and $p_{n+1}=1$ entail $p:\Pi p_{n+1}=p:\Pi p_n$.

Thus, if x_{n+1} is either a negative instance of H or a positive instance of A, then the fact that the law $H\subset A$ is automatically confirmed in its $n+1$:th instance does not increase its probability *a posteriori*. Paradoxical confirmations of a law, in other words, are ineffective from the point of view of probability.

A few words must be added concerning the Paradox of Confirmation in relation to the Frequency Interpretation of Inductive Probability.

ϕ_{n+1} means the set of remaining possible Necessary (Sufficient) Conditions of H (A) relative to x_1 and . . . and x_{n+1}.

If x_{n+1} is a negative instance of H, then x_{n+1} cannot exclude or eliminate any property from being a Necessary Condition of H.

Similarly, if x_{n+1} is a positive instance of A, then x_{n+1} cannot exclude or eliminate any property from being a Sufficient Condition of A.

Thus, if x_{n+1} is either a negative instance of H or a positive instance of A, then the remaining possible Necessary (Sufficient) Conditions of H (A) relative to x_1 and . . . and x_{n+1} are the same as the remaining possible Necessary (Sufficient) Conditions of H (A) relative to x_1 and . . . and x_n. The thing x_{n+1}, in other words, is wholly inefficient from the point of view of elimination.

As we already know, a thing which is inefficient from the point of view of elimination cannot affect the proportion which, on the Frequency Interpretation, answers to the probability *a posteriori* of the law.

The inefficiency of paradoxical confirmations of a law from the point of view of probability is thus but another expression for the inefficiency of paradoxical confirmations from the point of view of elimination.

The Paradox of Confirmation, therefore, is harmless.

6. *The Argument from Simplicity*

Another traditional idea of the Logic of Inductive Probability is that the probability (*a priori*) of a law is somehow proportionate to the law's simplicity. We shall call this idea the Argument from Simplicity.

It is usually not very clear what is meant by 'simplicity' as an attribute of laws. The idea of simplicity covers a wide range of cases.

There is, for instance, the case of the simplest curve connecting a number of points in a system of co-ordinates. The question of measuring degrees of simplicity in curves has been the object of much discussion. It will, however, not be treated in this book.

There is also the case when the simplicity of a law has something to do with the complexity of the conditioning relationship. This is the only case with which we shall deal here.

There are two forms of the Argument from Simplicity which at first sight seem to run contrary to one another. We shall call them the direct and the inverse form of the argument respectively. The first asserts that the less complex a property, the greater the probability that it will be a conditioning property of a given conditioned property. The second asserts that the more complex a property, the greater the probability that it will be a conditioning property of a given conditioned property. Thus the first form of the argument views Inductive Probability as directly, and the second as inversely, proportionate to simplicity in the laws under consideration.

We shall examine both forms of the argument within the framework of our Calculus of Probability. The results will immediately be presented and, as it were, 'mirrored' in the Frequency Interpretation.

A. The Direct Argument from Simplicity.

Two principal cases will be distinguished, according to whether the law is a Universal Implication or a Universal Equivalence.

Ai. Universal Implications.

Two sub-cases must be distinguished, according to whether we consider complexity in the antecedent or the consequent of the implication property which the law asserts to be universal.

Aia. Complexity in the antecedent.

Complexity means that the antecedent is a sum of other properties.

Let there be a sequence of properties $A_1, \ldots, A_n, \ldots$.

Let ϕ_1 denote the property of being a Necessary Condition of A_1.

Let ϕ_2 denote the property of being a Necessary Condition of $A_1 \mathrm{v} A_2$ or ΣA_2.

Similarly, we introduce ϕ_3, *etc.*

Let ϕ denote the property of being a Necessary Condition of *all A*-properties.

Every one of the members of the sequence $\phi_1, \ldots, \phi_n, \ldots$ is included in its predecessor. The logical product of *all* members is ϕ. Thus we have $lim(\phi_n, \phi)$.

Let ϕ_0 denote a set of properties from the same universe as the *A*-properties.

The following suppositions are introduced:

S1. The property H belongs to the set ϕ_0. Thus $E\,\phi_0$.

S2. There is a probability p that a random property, *e.g.*, H, will be a Necessary Condition of *all* the *A*-properties on the evidence that it belongs to ϕ_0. In symbols: $P(\phi, \phi_0, p)$.

S3. There is a probability p_1 that a random property, *e.g.*, H, will be a Necessary Condition of A_1 on the evidence that it belongs to ϕ_0. In symbols: $P(\phi_1, \phi_0, p)$. There is further a probability p_{n+1} that a random property, *e.g.*, H, is a Necessary Condition of ΣA_{n+1} on the evidence that it belongs to ϕ_0 *and* is a Necessary Condition of ΣA_n. In symbols:

$(n)P(\phi_{n+1}, \phi_0 \& \phi_n, p_{n+1})$.

S4. $p > 0$.

Given these four suppositions as data, we can deduce a probability that a random property, *e.g.*, H, is a Necessary Condition of ΣA_n on the evidence that it belongs to ϕ_0.

This deduction has, in fact, already been made. It is in all details exactly the same as the deduction which took us from the suppositions *S1–S4* of §3 of this chapter to the probability-expression $(n)P(\phi_n, \phi_0, \Pi p_n)$. (Cf. above p. 246 f.)

We also deduced $(n)(\Pi p_n > 0)$ and $lim(\Pi p_n, p)$. (Cf. above p.247f.)

Πp_n is the probability of the law $\Sigma A_n \subset H$ on the evidence that H belongs to ϕ_0.

If $p_{n+1} < 1$, then $\Pi p_{n+1} < \Pi p_n$.

Thus, provided it is not maximally probable that a random property H is a Necessary Condition of ΣA_{n+1} on the evidence that it belongs to ϕ_0 *and* is a Necessary Condition of ΣA_n, then the simpler law $\Sigma A_n \subset H$ is more probable than the law $\Sigma A_{n+1} \subset H$ on the evidence that H belongs to ϕ_0.

In the Frequency Interpretation this means:

Provided that it is not the case that practically all members of ϕ_0 which are Necessary Conditions of ΣA_n are also Necessary Conditions of ΣA_{n+1}, then it is more frequently the case that a random member of ϕ_0 is a Necessary Condition of ΣA_n than of ΣA_{n+1}.

This is a 'pure triviality.'

Aiβ. Complexity in the consequent.

Complexity means that the consequent is a product of other properties.

The reasoning is strictly 'dual' to the reasoning in the preceding case.

Let there be a sequence of properties $A_1, \ldots, A_n, \ldots$

Let ϕ_1 denote the property of being a Sufficient Condition of A_1.

Let ϕ_2 denote the property of being a Sufficient Condition of A_1 and A_2 or of ΠA_2. (Cf. above p. 41.)

Similarly, we introduce ϕ_3, *etc.*

Let ϕ denote the property of being a Sufficient Condition of *all A*-properties.

Every one of the members of the sequence $\phi_1, \ldots, \phi_n, \ldots$ is included in its predecessor member. Thus we have $lim(\phi_n, \phi)$.

Let ϕ_0 denote a set of properties from the same universe as the *A*-properties.

The following suppositions are introduced:

S1. The property H belongs to the set ϕ_0. Thus $E\,\phi_0$.

S2. There is a probability p that a random property, *e.g.*, H, will be a Sufficient Condition of *all* the *A*-properties on the evidence that it belongs to ϕ_0. In symbols: $P(\phi, \phi_0, p)$.

S3. There is a probability p_1 that a random property, *e.g.*, H, will be a Sufficient Condition of A_1 on the evidence that it belongs to ϕ_0. In symbols: $P(\phi_1, \phi_0, p)$. There is further a probability p_{n+1} that a random property, *e.g.*, H, will be a Sufficient Condition of ΠA_{n+1} on the evidence that it belongs to ϕ_0 *and* is a Sufficient Condition of ΠA_n. In symbols:

$(n)P(\phi_{n+1}, \phi_0 \& \phi_n, p_{n+1})$.

S4. $p > 0$.

Given these four suppositions as data, we can deduce $(n)P(\phi_n, \phi_0, \Pi p_n)$ and $(n)(\Pi p_n > 0)$ and $lim(\Pi p_n, p)$ in exactly the same way as those expressions were deduced from the suppositions $S1$–$S4$ of §3.

Πp_n is the probability of the law $H \subset \Pi A_n$ on the evidence that H belongs to ϕ_0.

If $p_{n+1} < 1$, then $\Pi p_{n+1} < \Pi p_n$.

Thus, provided it is not maximally probable that a random property H is a Sufficient Condition of ΠA_{n+1} on the evidence that it belongs to ϕ_0 *and* is a Sufficient Condition of ΠA_n, then the simpler law $H \subset \Pi A_n$ is more probable than the law $H \subset \Pi A_{n+1}$ on the evidence that H belongs to ϕ_0.

In the Frequency Interpretation this means:

Provided that it is not the case that practically all members of ϕ_0 which are Sufficient Conditions of ΠA_n are also Sufficient Conditions of ΠA_{n+1}, then it is more frequently the case that a random member of ϕ_0 is a Sufficient Condition of ΠA_n than of ΠA_{n+1}.

This again is a 'pure triviality.'

Aii. Universal Equivalences.

No proof as to the direct proportionality of simplicity and Inductive Probability is possible.

That such must be the case is immediately clear from the Frequency Interpretation. If H is a Necessary-and-Sufficient Condition of the sum ΣA_{n+1} or the product ΠA_{n+1} respectively, it does not follow that H will also be a Necessary-and-Sufficient Condition of the sum ΣA_n or the product ΠA_n respectively. (Cf. above Chap. III, §2 and Chap. IV, §8.) Consequently, the number of members of a set of properties ϕ_0 which are Necessary-and-Sufficient Conditions of ΣA_{n+1} or ΠA_{n+1} need not be either smaller than, or equal to, the number of members which are Necessary-and-Sufficient Conditions of ΣA_n or ΠA_n. It may be greater than that number.

Similar considerations apply to Necessary-and-Sufficient Conditions which are sums of products or products of sums of A-properties.

B. The Inverse Argument from Simplicity.

Two principal cases will again be distinguished, according to whether the law is a Universal Implication or a Universal Equivalence.

Bi. Universal Implications.

Two sub-cases must again be distinguished, according to whether we consider complexity in the antecedent or the consequent of the implication property which the law asserts to be universal.

Bia. Complexity in the antecedent.

Complexity means that the antecedent is a product of several properties.

Let there be a sequence of properties $A_1, \ldots, A_n, \ldots$.

Let ϕ_1 denote the property of being a Necessary Condition of A_1.

Let ϕ_2 denote the property of being a Necessary Condition of $A_1 \& A_2$ or ΠA_2.

Similarly, we introduce ϕ_3, *etc.*

Let ϕ denote the property of being a Necessary Condition of the product of *all* the A-properties.

Every one of the members of the sequence $\phi_1, \ldots, \phi_n, \ldots$ is included in its successor. The logical sum of *all* members is ϕ. Thus we have $lim(\phi_n, \phi)$.

Let ϕ_0 denote a set of properties from the same universe as the A-properties.

The following suppositions are introduced:

S1. The property H belongs to the set ϕ_0. Thus $E\ \phi_0$.

S2. There is a probability p that a random property, *e.g.*, H, will be a Necessary Condition of the product of *all* the A-properties on the evidence that it belongs to ϕ_0. In symbols: $P(\phi, \phi_0, p)$.

S3. There is a probability p_1 that a random property, *e.g.*, H, will be a Necessary Condition of A_1 on the evidence that it belongs to ϕ_0. In symbols: $P(\phi_1, \phi_0, p_1)$. There is further a probability p_{n+1} that a random property, *e.g.*, H, will be a Necessary Condition of ΠA_{n+1} but *not* of ΠA_n on the evidence that it belongs to ϕ_0. In symbols: $(n)P(\phi_{n+1} \& \overline{\phi_n}, \phi_0, p_{n+1})$.

Given the suppositions *S1* and *S3* as data, we can deduce a probability that a random property, *e.g.*, H, will be a Necessary Condition of ΠA_n on the evidence that it belongs to ϕ_0. Adding *S2* to the data, we can deduce that this probability approaches p as a limit.

If $n > 1$, then ϕ_n is identical with $\phi_1 \vee \Sigma(\phi_n \& \overline{\phi_{n-1}})$.

Any two members of the sequence of properties ϕ_1, $\phi_2\&\bar{\phi}_1$, . . ., $\phi_{n+1}\&\bar{\phi}_n$, . . . are mutually exclusive. In symbols: $(m)(n)(\overline{m=n}\rightarrow\bar{E}\ \phi_{m+1}\&\bar{\phi}_m\&\phi_{n+1}\&\bar{\phi}_n)$.

The Special Addition Principle and $E\ \phi_0$ and $(m)(n)(\overline{m=n}\rightarrow\bar{E}\ \phi_{m+1}\&\bar{\phi}_m\&\phi_{n+1}\&\bar{\phi}_n)$ and $P(\phi_1,\ \phi_0,\ p_1)$ and $(n)P(\phi_{n+1}\&\bar{\phi}_n,\ \phi_0,\ p_{n+1})$ entail $(n)P(\phi_n,\ \phi_0,\ \Sigma p_n)$.

If $p_{n+1}>0$, then $\Sigma p_{n+1}>\Sigma p_n$.

The Axiom of Continuity and $E\ \phi_0$ and $(n)P(\phi_n,\ \phi_0,\ \Sigma p_n)$ and $P(\phi,\ \phi_0,\ p)$ and $lim(\phi_n,\ \phi)$ entail $lim(\Sigma p_n,\ p)$.

Thus, provided it is not minimally probable that a random property, e.g., H, will be a Necessary Condition of ΠA_{n+1} but *not* of ΠA_n on the evidence that it belongs to ϕ_0, then the simpler law $\Pi A_n \subset H$ is less probable than the law $\Pi A_{n+1} \subset H$ on the evidence that H belongs to ϕ_0.

In the Frequency Interpretation this means:

Provided that it is not the case that practically no member of ϕ_0 is a Necessary Condition of ΠA_{n+1} but *not* of ΠA_n, then it is less frequently the case that a random member of ϕ_0 is a Necessary Condition of ΠA_n than of ΠA_{n+1}.

This again is a 'pure triviality.'

Biβ. Complexity in the consequent.

Complexity means that the consequent is a sum of other properties.

The reasoning is strictly 'dual' to the reasoning in the preceding case.

Let there be a sequence of properties $A_1, \ldots, A_n, \ldots$.

Let ϕ_1 denote the property of being a Sufficient Condition of A_1.

Let ϕ_2 denote the property of being a Sufficient Condition of $A_1 v A_2$ or ΣA_2.

Similarly, we introduce ϕ_3, *etc.*

Let ϕ denote the property of being a Sufficient Condition of the sum of *all* the A-properties.

Every one of the members of the sequence $\phi_1, \ldots, \phi_n, \ldots$ is included in its successor. The logical sum of *all* members is ϕ. Thus we have $lim(\phi_n,\ \phi)$.

Let ϕ_0 denote a set of properties from the same universe as the A-properties.

The following propositions are introduced:

S1. The property H belongs to the set ϕ_0. Thus $E \phi_0$.

S2. There is a probability p that a random property, *e.g.*, H, will be a Sufficient Condition of the sum of *all* the A-properties on the evidence that it belongs to ϕ_0. In symbols: $P(\phi, \phi_0, p)$.

S3. There is a probability p_1 that a random property, *e.g.*, H, will be a Sufficient Condition of A_1 on the evidence that it belongs to ϕ_0. In symbols: $P(\phi_1, \phi_0, p_1)$. There is further a probability p_{n+1} that a random property, *e.g.*, H, will be a Sufficient Condition of ΣA_{n+1} but *not* of ΣA_n on the evidence that it belongs to ϕ_0. In symbols: $(n)P(\phi_{n+1} \& \overline{\phi}_n, \phi_0, p_{n+1})$.

In exactly the same way as in the preceding case we deduce $(n)P(\phi_{n+1}, \phi_0, \Sigma p_n)$ and $lim(\Sigma p_n, p)$.

If $p_{n+1} > 0$, then $\Sigma p_{n+1} > \Sigma p_n$.

Thus, provided it is not minimally probable that a random property, *e.g.*, H, will be a Sufficient Condition of ΣA_{n+1} but *not* of ΣA_n on the evidence that it belongs to ϕ_0, then the simpler law $H \subset \Sigma A_n$ is less probable than the law $H \subset \Sigma A_{n+1}$ on the evidence that H belongs to ϕ_0.

In the Frequency Interpretation this means:

Provided that it is not the case that practically no member of ϕ_0 is a Sufficient Condition of ΣA_{n+1} but *not* of ΣA_n, then it is less frequently the case that a random member of ϕ_0 is a Sufficient Condition of ΣA_n than of ΣA_{n+1}.

This again is a 'pure triviality.'

Bii. Universal Equivalences.

No proof as to the inverse proportionality of simplicity and Inductive Probability is possible.

That such must be the case is immediately clear from the Frequency Interpretation. (Cf. above p. 260.)

*   *   *   *   *

No doubt the reconstructive examination of this section has led us to extremely trivial results. It seems to me, however, hardly reasonable to expect that anything much more interesting than this could emerge from a clarification of the vague ideas, entertained in science as well as in everyday life, concerning the relation of Inductive Probability to simplicity and complexity of logical conditions. The chief interest of the clarification is to show that we can assign even to these vague ideas a place within the framework of axiomatic probability.

7. *The Argument from Analogy*

Traditionally related to ideas on Inductive Probability is the Argument from Analogy:

Two things, x and y, are known to have n properties, A_1 and . . . and A_n, in common. The thing x is further known to possess the property H. It is thought that the probability that the thing y will also possess the property H is, in general, the greater, the larger the number n of properties which the two things are known to have in common.

Is this argument valid? If valid, what is its epistemological significance? The process of answering these questions we shall again refer to as a reconstructive examination of the argument.

It does not seem to me likely that a satisfactory answer to the above questions can be provided unless we conceive of the probability which is thought to increase with the analogy as the probability of a certain nomic connexion between the property H and the product of properties ΠA_n.

This attitude means that probability based on analogy is regarded as a kind of Inductive Probability. It means, moreover, that the Argument from Analogy is closely related to the Argument from Simplicity.

Three cases should be distinguished, according to whether H is viewed as Sufficient Condition or as Necessary Condition or as Necessary-and-Sufficient Condition of the product ΠA_n. Of these three cases, however, the first is hardly relevant to what is ordinarily called reasoning from analogy.

A. H is a Sufficient Condition of ΠA_n.

This case involves an application of variant $Ai\beta$ of the Argument from Simplicity.

Let there be a sequence of properties $A_1, \ldots, A_n, \ldots$.

The ϕ-properties are defined as in $Ai\beta$.

The four suppositions $S1–S4$ of $Ai\beta$ are introduced.

Given these four suppositions as data, we can deduce $(n)(E\,\phi_0\,\&\,\phi_n)$ and $(n)(\Pi p_n > 0)$ and $(n)P(\phi_n, \phi_0, \Pi p_n)$ and $lim(\Pi p_n, p)$

in exactly the same way as those expressions were deduced from the suppositions $S1$–$S4$ of §3.

ϕ is identical with $\phi \& \phi_n$.

Thus we have $(n)P(\phi \& \phi_n, \phi_0, p)$.

The Multiplication Principle and $(n)(E\ \phi_0 \& \phi_n)$ and $(n)P(\phi_n, \phi_0, \Pi p_n)$ and $(n)P(\phi \& \phi_n, \phi_0, p)$ and $(n)(\Pi p_n > 0)$ entail $(n)P(\phi, \phi_0 \& \phi_n, p : \Pi p_n)$.

$p > 0$ and $(n)(\Pi p_n > 0)$ entail $(n)(p : \Pi p_n > 0)$.

If $p_{n+1} < 1$, then $p : \Pi p_{n+1} > p : \Pi p_n$.

$p > 0$ and $lim(\Pi p_n, p)$ entail $lim(p : \Pi p_n, 1)$.

We can state what has been proved as follows:

Suppose that it is not minimally probable that a random property H will be a Sufficient Condition of *all* the A-properties on the evidence that it belongs to ϕ_0, and not maximally probable that a random property H will be a Sufficient Condition also of the $n+1$:th A-property, on the evidence that it belongs to ϕ_0 *and* is a Sufficient Condition of the first n A-properties. Then that a random property H will be a Sufficient Condition of *all* the A-properties is more probable on the evidence that it belongs to ϕ_0 *and* is a Sufficient Condition of the first $n+1$ A-properties, than on the evidence that it belongs to ϕ_0 *and* is a Sufficient Condition of the first n A-properties only. The increasing probability approaches 1 as its limit.

In the Frequency Interpretation this means:

Suppose that it is not the case that practically no property which is a member of ϕ_0 is a Sufficient Condition of *all* the A-properties, nor that practically all properties, which are members of ϕ_0 *and* Sufficient Conditions of the first n A-properties, are Sufficient Conditions of the $n+1$:th A-property also. Then it is more frequently the case that a random property H, which is a member of ϕ_0 *and* a Sufficient Condition of the first $n+1$ A-properties, is a Sufficient Condition of *all* the A-properties, than that a random property H, which is a member of ϕ_0 *and* a Sufficient Condition of the first n A-properties, is a Sufficient Condition of *all* the A-properties. n being indefinitely increased it tends to be practically always the case that a random property H, which is a member of ϕ_0 *and* a Sufficient Condition of the first n A-properties, is also a Sufficient Condition of *all* the A-properties.

B. H is a Necessary Condition of ΠA_n.

This case involves an application of variant *Bia* of the Argument from Simplicity.

Let there be a sequence of properties $A_1, \ldots, A_n, \ldots$.

The ϕ-properties are defined as in *Bia*.

The three suppositions *S1–S3* of *Bia* are introduced.

From these three suppositions as data we deduced $(n)P(\phi_n, \phi_0, \Sigma p_n)$ and $lim(\Sigma p_n, p)$.

We introduce a fourth supposition:

S4. $p_1 > 0$.

ϕ is identical with $\phi_1 v \phi \& \overline{\phi}_1$.

$\overline{E} \ \phi_1 \& \phi \& \overline{\phi}_1$ is tautologous.

The Special Addition Principle and $E \ \phi_0$ and $\overline{E} \ \phi_1 \& \phi \& \overline{\phi}_1$ and $P(\phi_1, \phi_0, p_1)$ and $P(\phi_1 v \phi \& \overline{\phi}_1, \phi_0, p)$ entail $P(\phi \& \overline{\phi}_1, \phi_0, p\text{-}p_1)$.

The Axiom of Minimum Probability and $E \ \phi_0$ and $P(\phi \& \overline{\phi}_1, \phi_0, p\text{-}p_1)$ entail $p\text{-}p_1 \geqslant 0$ which entails $p \geqslant p_1$.

$p \geqslant p_1$ and $p_1 > 0$ entail $p > 0$.

$\overline{E} \ \phi \& \phi_0$ entails $\phi_0 \subset \overline{\phi}$.

The Inclusion Axiom and the Addition Axiom and $E \ \phi_0$ and $\phi_0 \subset \overline{\phi}$ entail $P(\phi, \phi_0, 0)$.

The Axiom of Uniqueness and $E \ \phi_0$ and $P(\phi, \phi_0, p)$ and $P(\phi, \phi_0, 0)$ entail $p = 0$.

Thus, $p_1 > 0$ entails $E \ \phi \& \phi_0$.

ϕ_n is identical with $\phi \& \phi_n$.

The Multiplication Principle and $E \ \phi_0 \& \phi$ and $P(\phi, \phi_0, p)$ and $(n)P(\phi \& \phi_n, \phi_0, \Sigma p_n)$ and $p > 0$ entail $(n)P(\phi_n, \phi_0 \& \phi, \Sigma p_n : p)$.

If $p_{n+1} > 0$, then $\Sigma p_{n+1} : p > \Sigma p_n : p$.

$p > 0$ and $lim(\Sigma p_n, p)$ entail $lim(\Sigma p_n : p, 1)$.

We can state what has been proved as follows:

Suppose that it is not minimally probable that a random property *H* will be a Necessary Condition of the first *A*-property on the evidence that it is a member of ϕ_0, and not minimally probable that it will be a Necessary Condition of the product of the first $n+1$ *A*-properties without being a Necessary Condition of the product of the first n *A*-properties on the evidence that it is a member of ϕ_0. Then it is more probable that a random property *H* will be a Necessary Condition of the product of the first $n+1$ *A*-properties than of the product of the first n

A-properties, on the evidence that it is a member of ϕ_0 *and* a Necessary Condition of the product of *all* A-properties. The increasing probability approaches 1 as its limit.

In the Frequency Interpretation this means:

Suppose that it is not the case that practically no property which is a member of ϕ_0 is a Necessary Condition of the first A-property, nor that practically no property which is a member of ϕ_0 is a Necessary Condition of the product of the first $n+1$ A-properties without being a Necessary Condition for the product of the first n A-properties. Then it is more frequently the case that a random property H, which is a member of ϕ_0 *and* a Necessary Condition of the product of *all* the A-properties, is a Necessary Condition of the product of the first $n+1$ A-properties than a Necessary Condition of the product of the first n A-properties. n being indefinitely increased it tends to be practically always the case that a random property H, which is a member of ϕ_0 *and* a Necessary Condition of the product of *all* the A-properties, is a Necessary Condition of the product of the first n A-properties.

This case is relevant to reasoning from analogy. It means that if H is (known or assumed to be) necessary for the occurrence of *all* the A-properties in a thing, then, on suitable conditions, it becomes more and more probable, with increasing n, that H must be present on the occurrence of n A-properties in a thing.

C. H is a Necessary-and-Sufficient Condition of ΠA_n.

This case, like the first, involves an application of variant $Ai\beta$ of the Argument from Simplicity.

Whatever is a Necessary-and Sufficient Condition of ΠA_n is also a Sufficient Condition of ΠA_n.

ϕ_0 and the sequence $\phi_1, \ldots, \phi_n, \ldots$ mean the same as in $Ai\beta$.

ϕ denotes the property of being a Necessary-and-Sufficient Condition of ΠA_n.

The suppositions $S1$–$S4$ of $Ai\beta$ are introduced with the sole modification that in $S2$ we substitute 'Necessary-and-Sufficient Condition' for 'Sufficient Condition.'

In exactly the same way as before we can deduce
$(n)P(\phi, \phi_0 \& \phi_n, p : \Pi p_n)$ and $(n)(p : \Pi p_n > 0)$.

If $p_{n+1} < 1$, then $p : \Pi p_{n+1} > p : \Pi p_n$.

We cannot, however, in this case deduce $lim(\phi_n, \phi)$ which is necessary for the deduction of $lim(\Pi p_n, p)$ and $lim(p : \Pi p_n, 1)$ respectively. This is quite natural, considering that ϕ_n is a (decreasing) set of Sufficient Conditions and ϕ a set of Necessary-and-Sufficient Conditions.

We can state what has been proved as follows:

Suppose that it is not minimally probable that a random property H will be a Necessary-and-Sufficient Condition of the product of *all* the A-properties on the evidence that it is a member of ϕ_0, and not maximally probable that a random property H will be a Sufficient Condition of the $n+1$:th A-property also, on the evidence that it is a member of ϕ_0 *and* is a Sufficient Condition of the first n A-properties. Then, that a random property H will be a Necessary-and-Sufficient Condition of the product of *all* the A-properties is more probable on the evidence that it is a member of ϕ_0 *and* is a Sufficient Condition of the first $n+1$ A-properties, than on the evidence that it is a member of ϕ_0 *and* is a Sufficient Condition of the first n A-properties. The increasing probability cannot be proved to approach 1 as its limit.

In the Frequency Interpretation this means:

Suppose that it is not the case that practically no property which is a member of ϕ_0 is a Necessary-and-Sufficient Condition of the product of *all* the A-properties, nor that practically all properties, which are members of ϕ_0 *and* Sufficient Conditions of the first n A-properties, are Sufficient Conditions of the $n+1$:th A-property also. Then it is more frequently the case that a random property H, which is a member of ϕ_0 *and* a Sufficient Condition of the first $n+1$ A-properties, is a Necessary-and-Sufficient Condition of the product of *all* the A-properties, than that a random property H, which is a member of ϕ_0 *and* a Sufficient Condition of the first n A-properties, is a Necessary-and-Sufficient Condition of the product of *all* the A-properties. It cannot, however, be proved that what is here more frequently the case ends, in the long run, to become practically always the case.

This case is also relevant to reasoning from analogy. It means that if H is (known or assumed to be) sufficient for the occurrence of n A-properties in a thing, then, on suitable conditions, it becomes with increasing n more and more probable that H

is necessary as well as sufficient for the occurrence of *all* the *A*-properties in a thing.

Let us examine a concrete instance of reasoning from analogy in the light of the above abstract trends of thought.

Use of the Argument from Analogy is common in daily life, and well known, though not always well trusted, in science. In philosophy there is a famous use of the argument for proving, on probable grounds, the existence of other minds. We observe mental activity in ourselves introspectively, and extrospectively, the occurrence of bodily events, such as purposive behaviour or verbal and other symbolic acts. We observe in other persons similar bodily events. It is argued that the more similarity there is between those bodily events in myself and in other persons, the more probable it is that other persons' bodies also are associated with a mind.

I shall not here be at all concerned with the philosophic value and significance of the argument, which seem to me highly debatable. I shall only inquire what the argument *might* amount to in terms of our above reconstruction of reasoning from analogy.

Let mental activity be denoted by H and let the type of bodily changes under discussion be represented by the sequence $A_1, \ldots, A_n, \ldots$. The Argument from Analogy, when applied to the problem of other minds, obviously assumes the existence of some sort of connexion by law between the observed bodily phenomena and mental activity. The bodily events in me, we say, are 'due' or 'attributed' to the activity of my mind. The precise nature of this law-connexion, however, is usually not clearly stated. Three possibilities are at hand.

A. Mental activity is a Sufficient Condition of bodily behaviour.

If we do not regard it as *a priori* minimally probable that mental activity is sufficient for the production of a total $A_1, \ldots, A_n, \ldots$ of bodily changes, then *a posteriori*, *i.e.*, supposing we know that mental activity is sufficient to produce n of these changes, it is the more probable that mental activity is sufficient for the production of all the changes, the greater this number n is. The increasing probability here approaches the maximum value 1.

It is reasonable to think that mental activity cannot alone be sufficient for the production of macroscopic bodily changes. The mind has to work through a physical medium, the nervous system. Mental activity would thus be only a Contributory Condition. This point, however, is immaterial here, since we can conceive of the factor H of the argument as a complex factor, of which mind is a constituent only. Allowing for this, I think most people are willing to accept mental activity (volition) in them as a sufficient cause of part of their behaviour. (The philosophic or scientific significance, if any, of this popular idea will, as already observed above, not concern us here.)

The Argument from Analogy here amounts to saying that the more similarity there is between physical events of a certain type associated with my body and the same type of physical events in association with other persons' bodies, the more probable does it become that mental activity is sufficient for the production of bodily changes of the type in question.

It is clear that from this nothing at all can be concluded as to the probability that, since the observed bodily changes in my body are produced by mind in my case, they are also to be attributed to mental activity in the case of other persons. For the fact that H is a probable Sufficient Condition of something does not permit any probable conclusion from the occurrence of this 'something' to the occurrence of H, but only conversely. Thus from the fact that there is mental activity in me and the fact that mental activity is a probable Sufficient Condition of certain observed bodily changes in me *and* in other persons nothing can be concluded as to the probability that there is mental activity in those other persons too. (We only know that *if* there were mental activity in those persons it would probably be sufficient for the production of the observed changes in their bodies, but this is something entirely different from knowing that there probably is mental activity in them.)

The above means, generally speaking, that variant A of the Argument from Analogy has no direct bearing at all upon what is ordinarily understood by reasoning from analogy. (Cf. above p. 264.)

B. Mental activity is a Necessary Condition of bodily behaviour.

If we do not regard it as *a priori* minimally probable that mental activity is necessary for the production of a random (first) case of a variety A_1, . . ., A_n, . . . of bodily changes, then *a posteriori*, *i.e.*, supposing we know that mental activity is necessary for the production of the whole variety of changes, it is the more probable that mental activity is necessary for the production of n of those changes, the greater n is. The increasing probability approaches the maximum value 1.

If the range of behaviour A_1, . . ., A_n, . . . is not very narrow, it is fairly reasonable to think that no body is capable of behaving in all the ways under consideration unless it is associated with a mind. This premiss being accepted, the Argument from Analogy here amounts to saying that the more similarity there is between physical events of a certain type associated with my body and the same type of physical events associated with other persons' bodies, the more probable does it become that mental activity is necessary for the production of the similarity.

This is a genuine species of reasoning from analogy. It permits us to conclude with probability from the occurrence of mental activity in me and from the observation of similarities in the bodily life of myself *and* of other persons, to the occurrence of mental activity in those other persons too.

C. Mental activity is a Necessary-and-Sufficient Condition of bodily behaviour.

If we do not regard it as *a priori* minimally probable that mental activity is necessary as well as sufficient for the production of a total A_1, . . ., A_n, . . . of bodily changes, then *a posteriori*, *i.e.*, supposing we know that mental activity is sufficient for the production of n of these changes, it is the more probable that mental activity is necessary as well as sufficient for the production of all the changes, the greater this number n is. The increasing probability, however, cannot be proved to approach maximum probability.

As already observed, the supposition that mental activity is a sufficient cause of part of our bodily behaviour has a certain popular plausibility. This supposition being taken for granted, the Argument from Analogy here amounts to saying that the more similarity there is between physical events of a certain type associated with my body and the same type of events associated with other persons' bodies, the more probable does it become that mental activity is not only sufficient but also necessary for the production of bodily changes of the type in question.

This is also a relevant case of what is ordinarily understood by reasoning from analogy. The plausibility of the initial suppositions on which the argument is based appear to me stronger in case *C.* than in case *B.* The disadvantage of *C.* as compared with *B.* is that it cannot make the probability approach the maximum 1, unless we assume that everything which is sufficient for the production of all the bodily phenomena under discussion is also necessary for that purpose. Whether we are inclined to think such a supposition warranted in the case of the problem of other minds, I cannot say.

8. *Inductive Probability and Types of Law*

It frequently happens that we pass a judgment on the comparative probabilities of laws which is based, not on their confirmation by experience nor on their relative simplicity of structure, but on considerations as to the 'material' character or content of them. It is, *e.g.*, reasonable to assume that there are types of law in physics or the natural sciences which are commonly acknowledged to be more trustworthy than most types or any type of law in psychology or the social sciences.

The above ideas are very vague, and it is hardly to be expected that there is a unique way of reconstructing them. A first approach to their treatment may be suggested along the following lines:

Let H be a member of a set of properties ϕ, A of ψ, H' of ϕ', and A' of ψ'. The four properties ϕ, ψ, ϕ', ψ' can also be spoken of as the 'material types' of the four properties H, A, H', A'.

Let the laws, on the comparative probabilities of which we pass a judgment, be $H \subset A$ and $H' \subset A'$.

The proposition that $H \subset A$ is more probable than $H' \subset A'$ can be understood in a variety of different ways, of which we mention the following:

    i.   It is more probable that a property of the type ϕ has a

Necessary Condition of the type ψ than that a property of the type ϕ' has a Necessary Condition of the type ψ'. The property 'having a Necessary Condition of the type ψ' we denote by χ and the property 'having a Necessary Condition of the type ψ'' we denote by χ'. We have then two probability-expressions $P(\chi, \phi, p)$ and $P(\chi', \phi', p')$. It is asserted that $p > p'$.

ii. It is more probable that a property of the type ψ has a Sufficient Condition of the type ϕ than that a property of the type ψ' has a Sufficient Condition of the type ϕ'. The treatment is analogous to the case above.

One of the terms of the respective laws may also be taken as a constant property and not as the representative of properties of a certain 'material type.' We get two new interpretations:

iii. It is more probable that the property A will be a Necessary Condition of a property of the type ϕ than that A' will be a Necessary Condition of a property of the type ϕ'. We denote by χ the property 'Sufficient Condition of A' and by χ' the property 'Sufficient Condition of A'.' We have then two probability-expressions $P(\chi, \phi, p)$ and $P(\chi', \phi', p')$. It is asserted that $p > p'$.

iv. It is more probable that the property H will be a Sufficient Condition of a property of the type ψ than that H' will be a Sufficient Condition of a property of the type ψ'. The treatment is analogous to case iii.

A' may also be identical with A or H' with H. We then get a sub-case of iii and iv respectively. On these two alternatives it is necessary that the four properties should belong to one and the same Universe of Properties, but otherwise there is no objection to supposing that H and A belong to one and H' and A' to another universe.

Finally we can view H, A as a pair of properties from a set of pairs ϕ and H', A' as a pair from another set ϕ'. We then get a fifth interpretation:

    v. It is more probable that a property-pair of the type ϕ will constitute a condition-pair than that a property-pair of the type ϕ' will constitute a condition-pair.

If in i–v we replace 'more probable' with 'more frequently the case' we get the five ideas on the Frequency Interpretation. It must again be emphasized that the fact that laws cannot be verified through their instances and that consequently the frequencies of true laws cannot be determined even in finite sets, does not constitute any logical objection to the view that the comparative probabilities which we have been discussing are comparative truth-frequencies in sets of laws. The exemption from empirical control may make statements concerning such frequencies vague and useless in practice, but it does not render them logically absurd, nor does it alter the fact that people often believe them to be true.

I shall not discuss here how belief in propositions concerning the comparative probabilities of laws of different types arises. It is perhaps reasonable to think that we have come to entertain such beliefs from what confirmation and falsification of laws of the respective types has taught us, but it may sometimes also be the case that the beliefs are founded on pure intuition or vague analogies from other domains of our experience.

Chapter Ten

INDUCTION AND INVERSE
PROBABILITY

1. The Concept of Apparent Inductive Probability

IN Chap. VIII, §1 we distinguished between Real and Apparent Inductive Probability. The first is probability as an attribute of theories or laws, *i.e.*, of inductive conclusions of the second order. The second is probability as an attribute of predictions, *i.e.*, of inductive conclusions of the first order, which is for some reason mistakenly believed to be an attribute of theories or laws. As the most important sub-species of Apparent Inductive Probability we mentioned a number of traditional ideas concerning the relevance of probability to induction which may be grouped under the heading of Inverse Probability.

In Chap. IX we examined Real Inductive Probability. Some familiar arguments concerning the probability of laws were presented and reconstructively examined in the light of axiomatic probability and its interpretation as a frequency concept.

In the present chapter we shall deal with Apparent Inductive Probability. The treatment will be confined to Inverse Probability. The procedure is somewhat analogous to that of the foregoing chapter. Some main problems of Inductive Probability are presented and the 'traditional' arguments involving an application of Inverse Probability for their solution are outlined. Thereafter the problems are restated and examined in the light of axiomatic probability. The reconstructive examination serves two purposes. It shows that a certain pretended Inductive Probability is apparent only. And it provides the instrument for an estimation of the epistemological significance, if any, of the arguments examined.

There are three main cases in which Inverse Probability is traditionally thought relevant to induction. The first is known as the problem of the Probability of Causes. The second is the problem of evaluating, by means of the Inverse Principles of Maximum Probability and Great Numbers, the relevance of statistical samples to Statistical Laws. The third problem concerns the probability of 'future events.'

Each of the three problems exhibits a multitude of variants. Here we shall deal with the simplest cases only. This will be sufficient for the purpose of the philosophic criticism of Inverse Probability which we have in mind.

Be it remarked that the third of these problems, though it belongs to Inverse Probability, is not, strictly speaking, a problem in which we have to detect and unmask an Apparent Inductive Probability.

Be it also remarked that the use of these arguments in scientific reasoning belongs mainly to the past. They are to-day commonly regarded as logically 'unsound.' The philosophical criticism of them, however, has remained in part unaccomplished.

2. *The Probability of Causes*

These are problems and arguments which involve an application of the Inverse Principle of Probability. (Chap. VII, §10 and §13.)

As the prototype of such problems we may take the following:

$A_1, \ldots, A_n$ are the names of n possible 'causes' of an event B. The probabilities *a priori* that the respective causes will come to operate are $p_1, \ldots, p_n$. The probability or 'likelihood' that B will be produced, *if* $A_1, \ldots, A_n$ respectively operate, is $q_1, \ldots, q_n$ respectively. The event B has taken place. What is the probability *a posteriori* that B has been produced by a particular cause A_i? And which is the most probable cause?

The argument runs as follows:

The Inverse Principle supplies the value

$p_i q_i : (p_1 q_1 + \ldots + p_n q_n)$ in answer to the first question. The most probable cause, consequently, is that A_i for which $p_i q_i$, or the product of the probability *a priori* of the cause and the corresponding likelihood of the event, is greatest.

We now proceed to examine the problem and its solution in the light of axiomatic probability.

The proof of the Inverse Principle within the calculus rests on the assumption that what are here called the different 'causes' come to operate at least once within a certain realm of observations H.

The proof further assumes the different causes to be mutually exclusive.

Finally the proof assumes the disjunction of the different causes to be a Necessary Condition of a certain event B within the realm of observations H. The occurrence of the event, in other words, implies that one of the causes has been operating.

The above assumptions can be directly 'read off' from the proof of the Inverse Principle given in Chap. VII, §13. Relative to them the argument mentioned on the Probability of Causes is formally valid. Whether, however, the argument is relevant to inductive inference, can only be determined after it has been made clear to what kind of nomic connexion the word 'cause' refers.

It is evident that 'cause' cannot here imply Sufficient Condition. For then it would be absurd to say that the various causes produce the effect in question with various probabilities.

It is equally clear that 'cause' cannot imply Necessary Condition. For then it would be absurd to say that the effect renders, in an individual case of its occurrence, its various causes variously probable.

On the other hand, the reconstructive examination of the argument shows that the disjunction of the causes is a Necessary Condition of the observed event (in so far as it falls within the realm of observations H). From this it immediately follows that 'cause' here is actually understood to imply Substitutable Requirement. (This is a by no means uncommon use of the word 'cause' in ordinary language.)

The Substitutable Requirements are jointly exhaustive and mutually exclusive parts of a realm of observations $H\&B$. The problem with which we are dealing can now be restated as follows:

What is the probability that an observed occurrence x of a certain event B can be 'localized' in one of a number of mutually exclusive and jointly exhaustive parts $A_1, \ldots, A_n$ of a realm of observations $H\&B$? To this question the Inverse Principle gives an answer, provided that we know the probabilities $p_1, \ldots, p_n$ that a random member of the realm belongs to the respective parts, and the probabilities $q_1, \ldots, q_n$ that a random member of the respective parts is an occurrence of the event.

From the above we can draw two conclusions:

i. The argument from Probability of Causes is not an argument of Real Inductive Probability. The probability of the cause is not that of a nomic connexion between events, but of a feature of an individual occurrence of a certain event. This feature is the presence of a certain one of the event's Substitutable Requirements.

ii. The argument is significant only if the 'localization' of occurrences in different parts of a field of measurement is connected with difficulties, or as we may also say, if the 'identification' of the sub-fields is problematic. This is actually the case in certain games of chance which traditionally offer the best illustrations to problems of Probability of Causes and Inverse Probability in general. We can, for instance, think of H as the joint content of a number of urns of equal shape and size and colour, containing differently coloured balls; of $A_1, \ldots, A_n$ as the balls of the respective urns; and of B as the balls of a certain colour. Whether cases corresponding to this are found in nature is uncertain.

* * * * *

Of the many variants of the above problem one deserves special mention because of its historical importance. It is this:

An event B has taken place once, twice, $\ldots$, n times in succession. Is its occurrence due to 'chance' or is it produced by a 'cause'?

On the former alternative it is held to be as likely as not that the event will take place once, *i.e.*, the probability of a single occurrence of it is $1 : 2$. The occurrences are further supposed to be 'independent.' On the second alternative it is certain, or at least maximally probable, that the event will take place, provided the cause is operating. An easy application of the Multiplication Principle and the Inverse Principle yields the following answer to our question:

After one, two, . . ., n successive occurrences of the event it is more probable than not that the event is due to cause and not to chance, provided it is not *a priori* two or more than two, four or more than four, . . ., 2^n or more than 2^n times as probable that the event is due to chance than that it is due to cause. Thus, given the probabilities *a priori* of chance and cause respectively, we can tell exactly how many successive occurrences of the event are needed in order to make it more probable *a posteriori* that the event is due to cause than that it is due to chance.

We now proceed to the critical examination. In this we shall not question the condition of independence and the validity of the application of the Multiplication Principle involved.

The application of the Inverse Principle here rests on the three assumptions, that not everything that occurs within a certain realm of observations is due to chance, that the alternatives cause and chance are mutually exclusive, and that their disjunction is a Necessary Condition of one, two, . . ., n successive occurrences of a certain event within the realm H.

In this problem 'cause' *may* imply Sufficient Condition, though it will suffice to let it be a factor, the occurrence of which renders the occurrence of the event in question maximally probable.

For the same reason as in the first problem 'cause' *cannot* imply Necessary Condition. The disjunction, however, of cause and chance has to be a Necessary Condition of one, two, . . ., n occurrences of the event; from which it follows that cause and chance separately must be two Substitutable Requirements of one, two, . . ., n successive occurrences of the event within the realm H.

The two Substitutable Requirements, in other words, are mutually exclusive and jointly exhaustive parts of $H\&B$. It is assumed that the probability of the event in the one part is 1 and in the other part $1 : 2$. If the event occurs in the first part it is said to be due to cause, and if it occurs in the second part it is attributed to chance. The problem with which we are dealing can now be restated as follows:

Is it more probable that one, two, . . ., n successive occurrences of a certain event B can all be 'localized' within the first than within the second of two mutually exclusive and jointly exhaustive parts, A_1 and A_2, of a realm $H\&B$? The probability of a single occurrence of the event is 1 in A_1 and 1 : 2 in A_2. The occurrences of the event are independent. To this question the Multiplication Principle and the Inverse Principle give an answer, provided that we know the probabilities p_1 and p_2 that a random member of the realm H belongs to A_1 and A_2 respectively.

From this similar conclusions to i and ii above can again be drawn.—It is difficult to see how it could be established, except perhaps for a few exceptional cases, that in a field of measurement which includes a Sufficient Condition of a given property, the probability of the property in the remaining part of the field will be 1 : 2. Applications of the problem outside the realm of games of chance seem, therefore, to be practically out of the question.

3. Inductions from Statistical Samples

These are problems which involve an application of the Inverse Principles of Maximum Probability and/or Great Numbers.

As the prototype of such problems we may take the following:

We have examined n positive instances of a property H. Of them m have been found to be positive instances of the property A also. We have generalized that a proportion p of the property H is included in the property A. Can anything be concluded from the statistical sample to the probability of the Statistical Law?

Traditionally, the discussion has been mainly concerned with estimating the relevance of samples of the above type to propositions about the *probability* that a random H will be A. If, however, we interpret probability as frequency, the traditional way of treatment becomes immediately relevant to the problem of Inductive Probability raised above. The traditional argument can be outlined as follows:

Suppose that the occurrences of A (and not-A) in positive instances of H are independent events.[1] Then, if the probability that A is present in a random positive instance of H is p_i, the probability that A is present in m of n positive instances of H is $\binom{n}{m} \cdot p_i^m \cdot (1-p_i)^{n-m}$.

Suppose further that the probability *a priori* that p_i is the probability that A is present in a random positive instance of H is q_i. Let $p_1, \ldots, p_w$ be the possible values of the probability that A is present in a random positive instance of H. Then $q_1, \ldots, q_w$ are the corresponding probabilities *a priori*.

On the basis of these suppositions the Inverse Principle is used for calculating the probability *a posteriori*, i.e., relative to a sample of n positive instances of H containing m positive instances of A, that p_i is the probability that A is present in a random positive instance of H. For this probability we get the value $p_i^m \cdot (1-p_i)^{n-m} \cdot q_i : \sum_{\mu=1}^{w} p_\mu^m \cdot (1-p_\mu)^{n-m} \cdot q_\mu$.

If the possible p-values continuously cover the range from 0 to 1, if the index μ is itself the value of p_μ, and if q_μ is the probability *a priori* corresponding to p_μ, then the calculated probability assumes the form

$$p_i^m \cdot (1-p_i)^{n-m} \cdot q_i : \int_0^1 p_\mu^m \cdot (1-p_\mu)^{n-m} \cdot q_\mu \, d\mu.$$

Let there be j p-values $p_{i_1}, \ldots, p_{i_j}$ in the interval $m : n \pm \varepsilon$.

On this additional supposition the Extended Inverse Principle is used for calculating the probability *a posteriori*, i.e., relative to a sample of n positive instances of H containing m positive instances of A, that one of these j p-values is the probability that A is present in a random positive instance of H. For this probability we get the value

$\sum_{\mu=1}^{j} p_{i_\mu}^m \cdot (1-p_{i_\mu})^{n-m} \cdot q_{i_\mu} : \sum_{\mu=1}^{w} p_\mu^m \cdot (1-p_\mu)^{n-m} \cdot q_\mu$ and, on the supposition of a continuous distribution of p-values over the interval from

---

[1] In order to express the condition of independence in the language of our calculus, we have to replace A by a sequence $A_1, \ldots, A_n, \ldots$ which can be interpreted as meaning: the occurrence of A in the first thing, $\ldots$, the occurrence of A in the n:th thing, $\ldots$. (Cf. above p. 211.)

o to 1, the value $\int_{m:n-\epsilon}^{m:n+\epsilon} p_\mu{}^m \cdot (1-p_\mu)^{n-m} \cdot q_\mu \, d\mu : \int_{0}^{1} p_\mu{}^m \cdot (1-p_\mu)^{n-m} \cdot q_\mu \, d\mu$.

If all probabilities *a priori* are equal, then according to the Inverse Principle of Maximum Probability the most probable value of p is that which comes closest to $m:n$. On the Frequency Interpretation this means that a sample of n things of which m are positive instances of A is most frequently a sample from a field of which the Statistical Law is true that a proportion $m : n$ of members of the field are positive instances of A.

If the sums $\sum_{\mu=1}^{j} q_{i_\mu}$ and $\int_{m:n-\epsilon}^{m:n+\epsilon} q_\mu \, d\mu$ respectively of probabilities *a priori* are not o, then, according to the Inverse Principle of Great Numbers, the probability that the value of p which comes closest to $m:n$ is the probability that A is present in a random positive instance of H increases with n and approaches the maximum value 1 as n is indefinitely increased. On the Frequency Interpretation this means that an ever-increasing sample of n things of which m are positive instances of A, is more and more frequently and finally 'almost always' a sample from a field of which the Statistical Law is true that a proportion $m:n$ of members of the field are positive instances of A.

We now proceed to an examination of the problem and its solution in the light of axiomatic probability and its interpretation as a frequency concept.—In this we shall disregard the assumption of independence and concentrate exclusively on a discussion of the probabilities *a priori* and *a posteriori* of the argument.

From the treatment in Chap. VII, §16 we know that the probabilities *a priori* are the probabilities that a random member of a field of measurement is also a member of one of a number of mutually exclusive parts of that field. Each of these parts of the field is characterized by a certain probability that a random member of the part is a positive instance of a given property.

From this immediately follows an important conclusion which the traditional argument tends to obscure. It is not sufficient for a correct treatment of the problem to take into consideration statistical samples of *one* property H only. The property H of our problem is but one of a number of mutually exclusive properties which jointly make up a large range of phenomena. Each of these properties is characterized by a certain probability that a random positive instance of it is also a positive instance of an other given property. On the Frequency Interpretation this means that each property is characterized by a certain statistical correlation with one and the same given property.

The assumption that the probabilities *a priori* are all equal would mean, on the Frequency Interpretation, that a random member of a certain field of measurement (the large range of phenomena) is equally as often a member of one (*e.g.*, *H*) of a number of mutually exclusive parts of that field as of any other part.

It is important to observe that the assumption of equal probabilities *a priori* cannot, on the Frequency Interpretation, be supported by reference to the alleged fact (hardly true) that the proportions occurring in nature actually have one value as often as any other.[1] For, in the first place we are not concerned with proportions occurring in nature as such, but with proportions within (mutually exclusive) parts of a main field of measurement, the nature of which has to be specified in each concrete case of application. And in the second place the probabilities *a priori* do not indicate the statistical distribution of those proportions over the sub-fields, but the distribution of members of the main field over the sub-fields of which those proportions are characteristic. This attempt to support the assumption of equal probabilities *a priori* is thus guilty of two grave errors, of which the second is already known to us from the exposition in Chap. VII, §17 as a tendency to misinterpret the probabilities *a priori* of the problem as probabilities of a higher 'type' than the other probabilities involved.

After this clarification of the nature of the probabilities *a priori* we can give a correct re-statement of the problem of the probabilities *a posteriori* as follows:

What is the probability that a statistical sample of *n* things containing *m* positive instances of the property *A* can be

---

[1] Cf. Edgeworth, *The Philosophy of Chance* in Mind 9 (1884): 'The assumption that any probability-constant about which we know nothing in particular is as likely to have one value as another is grounded upon the rough but solid experience that such constants do, as a matter of fact, as often have one value as another.'

'localized' in one of a number of mutually exclusive sub-fields of a main field of observations? Each of the sub-fields is charac-, terized by a probability that a random member of it is a positive instance of A (the p-values), and also a probability that a random member of the main field is a positive instance of the sub-field (the q-values).

To this question the Inverse Principle gives an answer. Provided the q-values are all equal, the Inverse Principle of Maximum Probability tells us that the sample can most probably be localized in a sub-field with a p-value as close as possible to $m:n$, or, on the Frequency Interpretation, in a sub-field of which the Statistical Law is true that a proportion as close as possible to $m:n$ of its members are positive instances of A. Provided the sum of q-values in an interval round $m:n$ is not 0, the Inverse Principle of Great Numbers tells us that the localization of the sample in a sub-field with a p-value in this interval is the more probable, the greater n is, and becomes maximally probable as n approaches infinity.

From this reconstructive examination of the problem and its solution two conclusions can be drawn:

i.   The argument examined cannot be one of Real Inductive Probability. For the calculated probability is, on the Frequency Interpretation, not the probability that a Statistical Law is true, but the probability that a statistical sample is from a field of observations, *of which* a certain Statistical Law is already assumed to be true.

ii. The argument is significant only if the 'localization' of samples in different parts of a main field of observations is connected with difficulties or, as we may also say, if the 'identification' of the sub-fields is problematic.

An adequate illustration of the problem and the argument for the case when there are a finite number of p-values, may be found in games of chance.—We can, *e.g.*, think of the main field as (throws with) the dice of an urn. The sub-fields are (throws with) those dice which are characterized by a certain probability of getting a one. The p-values are the probabilities of a one with the respective dice. The q-values are the probabilities that we are throwing with a die of a certain sub-field. On the Frequency Interpretation the p-values are the proportions of ones among throws with the respective dice, and the q-values are the proportions of throws with a die of the respective sub-fields, among throws with a die of the main field.

In n throws with a certain die we get a one m times. This information can now be used for estimating the probability that the die with which we have been throwing is characterized by a certain probability of getting a one. On the Frequency Interpretation: we can calculate the proportion, among all samples of m ones in n throws with any die, of samples obtained by throwing with a die of which a certain proportion of ones is characteristic.

It is clear that this estimate is significant only if it is not possible to identify directly the various dice as members of the respective sub-fields. If, *e.g.*, in determining the p-values, each die with the same p-value were painted the same colour or the p-value were just written on the die's surface, then we would use the colour or the written mark for the purpose of subsequently identifying a chosen die as a member of a certain sub-field. It would be of no use to produce a statistical sample and resort to Inverse Probability for the purpose of identification.

An adequate illustration of the problem and the argument for the case when the p-values continuously cover the range from 0 to 1, would have to make use of a non-denumerable set of things. No such picture is possible in 'empirical' realms, for which reason this case may be regarded as purely 'fictitious.'

The use of the Inverse Principles of Maximum Probability and Great Numbers for purposes of statistics depends upon whether there occur in 'nature' cases which are in all relevant features analogous to the above illustration from games of chance. Whether there are such cases or not, I am not competent to judge. That their range must be very limited as compared with the range of cases to which the principles of Inverse Probability were originally thought to be applicable, is quite clear.

It may be finally remarked that, on the Frequency Interpretation, any estimation of probabilities in virtue of the Inverse

Principles of Maximum Probability and Great Numbers depends, for its truth, on the truth of the Statistical Laws asserting the respective p- and q-values. The estimate of probability does not give us any additional reasons whatever for believing these laws to be true. This fact would in any case minimize the practical importance of Inverse Probability for statistical purposes.

4. The Probability of Future Events

These are problems which involve an application of the Principle of Succession (and similar rules).

As the prototype of such problems we may take the following:

We have examined n positive instances of a property H. All of them have been found to be positive instances of the property A also. Can anything be concluded from this as to the probability that the next positive instance of H will be a positive instance of A?

Suppose that the occurrences of the property A (and not-A) in positive instances of H are independent events. Then, if the probability that A is present in a random positive instance of H is p_i, the probability that A is present in n positive instances of H is p_i^n and in $n+1$ instances is p_i^{n+1}.

Let $p_1, \ldots, p_w$ be the possible values of the probability that A is present in a random positive instance of H. Let $q_1, \ldots, q_w$ be the corresponding probabilities a priori.

On the basis of these suppositions we can make a combined use of the Inverse Principle, the Multiplication Principle, and the Addition Principle for computing the probability that the next positive instance of H will be a positive instance of A, if the n first instances have been so. This probability is

$$\sum_{\mu=1}^{w} p_\mu^{n+1} \cdot q_\mu : \sum_{\mu=1}^{w} p_\mu^n \cdot q_\mu.$$

In virtue of the Principle of Succession the calculated probability increases with n. The probability, in other words, that the next positive instance of H will be a positive instance of A is greater, the greater the number of positive instances of H which are already known to be positive instances of A.

On the additional assumptions that all the q-values are equal and that the p-values continuously cover the range from 0 to 1, the calculated probability assumes the form $\int_0^1 p^{n+1} \, dp : \int_0^1 p^n \, dp$. After integration we get from this the famous value $(n+1) : (n+2)$.

The reconstructive examination of the problem and its solution is in all essentials similar to the examination of the argument of Inverse Probability in the preceding section.

In particular, the crucial probabilities *a priori* mean the same in both the cases. After having clarified their nature we can restate the problem about the probability to be calculated as follows:

We have a sample of n things from one of a number of mutually exclusive sub-fields of a main field of observations. Each of the sub-fields is characterized by a certain probability (the p-value) that a random member of it is a positive instance of a property A. We also know the probabilities (the q-values) that a random member of the main field belongs to a certain sub-field. All things in the sample are positive instances of A, but we do not know from which sub-field the sample is. What is the probability that, if the sample is enlarged by the incorporation in it of one new thing, this new thing is also a positive instance of A?

From this restatement of the problem this important conclusion can be drawn:

The whole argument is significant only if the 'localization' of samples in different parts of the main field is connected with difficulties, *i.e.*, if the 'identification' of the sub-fields is problematic. For, with regard to each of the sub-fields the probability that a random member of it will have the property A is assumed to be known. Therefore, if it were known from which sub-field the sample is taken, we should use this information for ascertaining the probability that the next thing will have the property A and the calculation in virtue of principles of Inverse Probability would become useless.

Adequate illustrations of the legitimate use of the argument may again be found in games of chance. *E.g.*, the illustration from the preceding section could be used.

Applications of the Principle of Succession outside the realm of games of chance depend upon whether there occur in 'nature' cases which are in all relevant features analogous to the case mentioned in the preceding section of dice in an urn.

$$* \qquad * \qquad * \qquad * \qquad *$$

Traditional use of the Principle of Succession for inductive purposes has, at least in most cases, been definitely a misuse. Under this head comes, *e.g.*, its use for computing the probability that the sun will rise to-morrow on the evidence that the sun has regularly risen every morning during so and so many years in the past, or that the next raven will be black on the evidence that all ravens observed in the past have had this colour.

Let us show this in some detail by examining what it would come to that, n being the number of observed black ravens, $(n+1):(n+2)$ or $\sum_{\mu=1}^{w} p_{\mu}^{n+1}.q_{\mu} \; \sum_{\mu=1}^{w} p_{\mu}^{n}.q_{\mu}$ is the probability that the next raven to be observed will also be black.

First it is necessary that a certain number of assumptions as to the constitution of things should be made in order that a correct deduction of the two values may become possible. We shall here ignore the assumptions of independence which are needed and only mention the following three suppositions:

i. There must be a set of fields of measurement for the probability that a thing is black. It would here be plausible to think of this set or this main field as the totality of species of birds. In order to obtain, without restrictions, the value $(n+1):(n+2)$ it must further be assumed that the number of members in the set, *e.g.*, the number of species of birds, is non-denumerably great. This fictitious assumption can be dropped, if we content ourselves with the value $\sum_{\mu=1}^{w} p_{\mu}^{n+1}.q_{\mu} : \sum_{\mu=1}^{w} p_{\mu}^{n}.q_{\mu}$ instead of $(n+1):(n+2)$.

ii. We must be given the probabilities that random members of those fields, *e.g.*, individual birds of the respective species, are black. Since the various species of birds seem to be, on the whole, associated with specific and constant combinations of colours, it is highly plausible to think that the probabilities in question will mainly be one of the extreme values 0 or 1 or some value close to them. In order to obtain the value $(n+1):(n+2)$ it must, however, be assumed that all intermediate probabilities will also be represented. This assumption can again be dropped if we content ourselves with the value $\sum\limits_{\mu=1}^{w} p_\mu{}^{n+1}.q_\mu : \sum\limits_{\mu=1}^{w} p_\mu{}^{n}.q_\mu.$

iii. Finally we must be given the probabilities *a priori* that random things, *e.g.*, randomly chosen individual birds, are members of a field, *e.g.*, belong to a species associated with a certain probability of black colour. In order to obtain the value $(n+1):(n+2)$, but not if we content ourselves with $\sum\limits_{\mu=1}^{w} p_\mu{}^{n+1}.q_\mu : \sum\limits_{\mu=1}^{w} p_\mu{}^{n}.q_\mu,$ we must assume all these probabilities to be equal. The practical absurdity of this assumption is quite obvious from what was said above about the colours of birds!

On the above basis we can estimate the probability that the next individual bird of a random species, of which n black individuals have been successively observed, will also be black.

Quite apart, however, from all considerations as to the practical plausibility or absurdity of conditions i–iii there is a deeper reason which renders futile not only the calculations, leading to the value $(n+1):(n+2)$, but also those much more modest ones leading to $\sum\limits_{\mu=1}^{w} p_\mu{}^{n+1}.q_\mu : \sum\limits_{\mu=1}^{w} p_\mu{}^{n}.q_\mu.$ This reason is that we know the observed birds to be *ravens*, *i.e.*, that we have already 'localized' the observed birds in a certain species or, as we may also say, that we have 'identified' the birds as ravens. Now the calculations presuppose that we have estimated the probabilities of black colour in random individuals of the respective species. (In the case of ravens it is plausible to think that this probability has been estimated as 1.) Thus, if the species is known, the estimated probability of blackness associated with *it*, and not the calculated probability of blackness associated with a sample of n black birds of *any* species, would be relevant to the question of the colour of the next bird of the species.

In a similar manner we can show the inapplicability of the Principle of Succession to any events which, like sunrises or being a raven, can be readily 'identified' as members of a certain one of a number of different fields of measurement of probabilities.

5. Remarks on the Historical Development of the Logic of Inductive Probability

From the general nature of probability as a substitute for certainty (p. 223) it follows that the history of probability is closely connected with the history of the philosophical and logical problems of induction. We shall, however, disregard this connexion here in so far as probability as a mere attribute of predictions is concerned. Our treatment will be confined to probability as a supposed or real attribute of laws.

In the history of this subject two trends can easily be distinguished. The first is mainly concerned with developing the philosophical implications of what is traditionally called *a posteriori* or Inverse Probability. It has been largely characteristic of this trend that it has confused what is really the probability of a prediction with the probability of a law, of which the prediction is a test-condition. The first trend, therefore, is essentially the history of what we have called Apparent Inductive Probability. The second is the history of Real Inductive Probability.

*　　*　　*　　*　　*

It is a remarkable fact which, it seems to me, has not been sufficiently noted by historians, that James Bernoulli in proving his famous theorem was really intending to solve a problem of Inverse Probability. This is quite plain from the fourth chapter of the fourth book of the *Ars Conjectandi*, where the author makes a distinction between the determination of probabilities *a priori*, *i.e.*, on the basis of an enumeration of possible and favourable alternatives, and *a posteriori*, *i.e.*, on the basis of statistical samples. He mentions a principle for the *a posteriori* determination of probabilities which comes very close to what we have called the Inverse Principle of Great Numbers. Of this principle he says that he has proved it after twenty years of effort and that the proof will be given in the work. Then follows the fifth chapter. Here Bernoulli proves what we have called the Direct Principle of Great Numbers and the book suddenly comes to an end. The *Ars Conjectandi* is unfinished, the author's work being interrupted by his death in 1705. It was edited posthumously by Nicholas (I) Bernoulli in 1713.

De Moivre also, who in his *Doctrine of Chances* (1716) contributed to the mathematical refinement of Bernoulli's theorem, was interested in Inverse Probability. He did not, however, confuse the Direct and the Inverse Principles of Great Numbers as Bernoulli himself most probably had done. But he believed the inverted principle to follow from the direct by a simple conversion. He says[1]: 'As, upon the Supposition of a certain determinate Law according to which any Event is to happen, we demonstrate that the Ratio of Happenings will continually approach to that Law, as the Experiments or Observations are multiplied: so, *conversely*, if from numberless Observations we find the Ratio of the Events to converge to a determinate quantity, as to the Ratio of P to Q; then we conclude that this Ratio expresses the determinate Law according to which the Event is to happen. For let the Law be expressed not by the Ratio $P : Q$, but by some other, as $R : S$; then would the Ratio of the Events converge to this last, not to the former: which contradicts our *Hypothesis*.' From this interesting quotation it is seen that de Moivre confused maximum probability with certainty. His reasoning would have been correct if Bernoulli had proved that, supposing an event's probability to be p, the *proportion* of that event's happening on all occasions is (certainly) p. But what Bernoulli really proved was only that the *probability* of the event's happening in a certain proportion, close to p, of all cases in a sample, increases with the size of the sample and approaches maximum probability 1 as the sample is indefinitely enlarged.

Further early approaches to the inverse problems were made

[1] *Op. cit.*, 3rd Ed. (1756), p. 251.

in several writings by Daniel Bernoulli in the period from 1734 to 1777. Of most interest from the point of view of the history of ideas is probably his *Essai d'une nouvelle analyse de la mortalité causée par la petite vérole* of the year 1760.

The first, however, to give a satisfactory foundation of Inverse Probability in the Calculus was Thomas Bayes in his *Essay Towards Solving a Problem in the Doctrine of Chances*, posthumously communicated by Richard Price for publication in the transactions of the Royal Society in 1763. Bayes proved what is called by us the Inverse Principle of Maximum Probability. He did not, however, prove the Inverse Principle of Great Numbers which, as we have seen, had been aimed at by various authors ever since the days of James Bernoulli. Bayes's paper is a masterpiece of mathematical elegance and free from the obscure philosophical pretensions which were soon to become associated with his achievement.

These pretensions are predominant already in the preface and long appendix which Price added to Bayes's original essay. Price talks eloquently of the importance of Bayes's discoveries for estimating the probability of causes and of inductive conclusions in general. He also tried to extend the purely mathematical achievements of Bayes. He stated the Inverse Principle of Great Numbers and a Principle of Succession, but he does not give very satisfactory proofs.

It appears that Laplace was not acquainted with the work of Bayes and Price when he developed the fundamentals of Inverse Probability in his *Mémoire sur la probabilité des causes par les événements* of the year 1774. Here he proved not only the Inverse Principle and the Inverse Principle of Maximum Probability, but also the Inverse Principle of Great Numbers and the Principle of Succession—on the simplifying assumption of equal probabilities *a priori*. In later writings, however, Laplace pays due attention to the work of his predecessors.

The mathematical treatment of Inverse Probability in Laplace is intimately connected with philosophical claims and hopes as regards the importance of the Calculus of Probability. Laplace is the founder of a mighty tradition in the history of scientific ideas, the last traces of which have survived till recent times and in which Inverse Probability holds a dominant position. We need not here follow its course in any detail. One of its foremost champions in England was de Morgan. The faith which he puts in Inverse Probability in general and in the formula $(n+1):(n+2)$ for the determination of the probability of future events in particular, cannot but amaze a modern reader by its complete lack of self-criticism. The same is largely true also of the use which Jevons made of Inverse Probability.[1]

First among the critics of Inverse Probability we should mention Leibniz, in his correspondence with James Bernoulli. Leibniz took a very sceptical attitude as regards that faith in Inverse Probability which Bernoulli thought he had already justified by the proof of his theorem. Leibniz acutely observed that any use of the calculus for attributing probability to inductive conclusions must rest upon initial assumptions which are themselves of an inductive character.[2] It is interesting to note that Laplace[3] and adherents of his school of thought[4] were also incidentally aware of this, without, however, realizing its significance for the philosophic claims which they associated with their treatment of Inverse Probability.

A guarded attitude as regards the traditional uncritical employment of Inverse Probability is noticeable in Leslie Ellis and Boole. The notorious Principle of Succession was most acutely criticized by Pierce. Under the influence of the renewal of the frequency view of probability and, in particular, thanks

[1] A characteristic use of the formula for determining with probability whether a phenomenon is due to a cause or to chance (cf. above p. 278 ff.) was made by Kirchhoff in the famous essay *Untersuchungen über das Sonnenspektrum* (1861). Kirchhoff, by this use of Inverse Probability, calculated the probability that the occurrence of sixty dark rays, observed by him at characteristic places in the spectrum of the sun, would coincide by chance with the rays in the spectrum of iron.— Deplorable uncritical use of the formulae for Probability of Causes are to be found in Hartmann's *Philosophie des Unbewussten* (1869). Hartmann, *inter alia*, calculates the probability for a non-material cause operating in a given case to be 0.9999985!

[2] See Leibniz's *Mathematische Schriften*, ed. by Gerhardt (1849–63), Vol. III. pp. 79–98 for his correspondence with James Bernoulli. Leibniz (letter of 3rd Dec., 1703), in opposition to the idea of Bernoulli that one could determine the probable value of human life with ever-increasing probability on the basis of statistical samples, makes the very acute observation that *'novi morbi inundant subinde humanum genus, quodsi ergo de mortibus quotcunque experimenta feceris, non ideo naturae rerum limites posuisti, ut pro futuro variare non possit.'* Bernoulli (letter of 20th April, 1704) was wholly unable to grasp the significance of this profound remark.

[3] *Essai philosophique sur les probabilités* (reprinted in Œuvres completes 1891–8), pp. xiv, xlviii, and liii, *etc.*

[4] Cf. Condorcet, *Essai sur l'application de l'analyse à la probabilité des décisions rendues à la pluralité des voix* (1785), p.x.

to the rapid development of the science of statistics in the last fifty years, Inverse Probability has gradually been losing ground. Foremost among its critics in modern times ranks R. A. Fisher, who emphatically rejects[1] it altogether.

In spite of the high merits of Fisher's work in theoretical statistics and in spite of the undeniable soundness of his practical attitude, his criticism of Inverse Probability remains to me very obscure and unsatisfactory from the point of view of logic.

Fisher[2] gives three main reasons for the rejection of Inverse Probability, *viz.*

   i.   that the Axiom of Bayes leads to contradictions,
  ii.   that the Axiom of Bayes is not a necessary truth, and
 iii.   that Inverse Probability cannot justify induction.

By the 'Axiom of Bayes' Fisher does not mean any of the inverse principles or formulae of our calculus. Exactly what he means by it is not quite clear to me, but I think he has in mind the distribution of equal probabilities *a priori* in accordance with the rule commonly known as the Principle of Non-Sufficient Reason or Principle of Indifference. This rule was actually used by Bayes for securing the equality of the probabilities *a priori* which is essential to the Inverse Principle of Maximum Probability (though not to some other inverse theorems).

We need not here discuss this famous principle. The charge that its use leads to contradictions is, I think, preposterous and must be withdrawn, if due attention is paid to the fact that probability must be taken relative to some field of measurement. This, however, is of minor importance. What matters is that Fisher's rejection of Inverse Probability does not, so far as I can see, involve any charge against the correctness, from the point of view of logic (mathematics), of the various Inverse Principles of the Calculus of Probability.

The essential point of Fisher's criticism, as I understand it, is the contention that Inverse Probability must be rejected

---

[1] *Statistical Methods for Research Workers* (1925), p. 10: 'The theory of inverse probability is founded upon an error, and must be wholly rejected.'
[2] *The Design of Experiments* (3rd Ed. 1942), p. 6 f.

because of its uselessness for practical purposes. This contention, no doubt, is largely true, but *why* it is has not been satisfactorily shown by Fisher.

As I have endeavoured to show in this book, the possible range and relevance of applications of Inverse Probability can be seen in a clearer light when the principles and formulae are given a solid foundation in axiomatic probability. The range is found to be narrowly restricted to cases which are analogous to some cases in games of chance. The overrated importance of Inverse Probability is due to a failure to note these conditions of analogy and to an overhasty assimilation to patterns from games of chance of cases which are essentially different from them. Fisher, therefore, is probably right in saying that most use of Inverse Probability (outside games of chance) has to be rejected. But he would be mistaken in claiming that Inverse Probability cannot retain its position as a logically creditable part of the Calculus of Probability.

*　*　*　*　*

The rationalistic era which saw the birth of modern natural science was much inclined to underrate (in spite of Bacon) the scientific importance of induction at the cost of deduction. Thus, *e.g.*, Galileo was well aware of the impossibility of achieving certainty by means of induction,[1] but nevertheless considered the mathematical laws of nature discovered through his 'resolutive' and 'compositive' methods to be absolute truths.[2] It was not until the time of Huyghens and Newton that it began gradually to become clear that even mathematical laws of nature, in so far as they possess material content, are inductive conclusions and as such only 'probably' true.

'The arguing from experiments and observations by induction,' says Newton in his *Opticks* (1704), is 'no demonstration of general conclusions.' And in the preface to his *Traité de la lumière* (1690) Huyghens writes:

'On y verra de ces sortes de demonstrations, qui ne produisent pas une certitude aussi grande que celles de Geometrie, et qui mesme en different beaucoup, puisque au lieu que les Geometres

---

[1] Against Vincenzo di Grazia in *Opere Complete* (ed. Albèri), Vol. XII, p. 513.
[2] *Dialoghi delle nuove scienze* in *Opere Complete* (ed. Albèri), Vol. XIII, p. 166, 171 and *passim*.

prouvent leurs Propositions par des Principes certains et incontestables, icy les Principes se verifient par les conclusions qu'on en tire; la nature de ces choses ne souffrant pas que cela se fasse autrement. Il est possible toutefois d'y arriver à un degré de vraisemblance, qui bien souvent ne cede guere à une evidence entiere. Sçavoir lors que les choses, qu'on a demon-trées par ces Principes suppozes, se raportent parfaitement aux phenomenes que l'experience a fait remarquer; sur tout quand il y en a grand nombre, et encore principalement quand on se forme et prevoit des phenomenes nouveaux, qui doivent suivre des hypotheses qu'on employe, et qu'on trouve qu'en cela l'effet repond à nostre attente.'

As will be seen, this quotation contains an allusion to an argument of Inductive Probability, *viz.*, the Argument from Confirmation. At roughly the same time as Huyghens, Leibniz made use of another argument of Inductive Probability, *viz.*, the Argument from Simplicity, in his famous comparison between discovering laws of nature and solving a cryptogram.[1]

It was, however, long before there were any attempts at a systematic treatment of (Real) Inductive Probability.

In his *System der Logik* (1811) Fries distinguished between mathematical and philosophical probability. The distinction was further developed by Fries himself in his *Kritik der Prinzipien der Wahrscheinlichkeitsrechnung* (1842) and adopted by Cournot in his *Exposition de la théorie des chances et des probabilités* (1843). It occurs in a great number of logicians and philosophers (Beneke, Drobisch, *etc.*) of the nineteenth century.

By philosophical probability those authors understood, broadly speaking, what we have called (Real) Inductive Probability. An essential difference between philosophical and mathematical probability, according to them, was that the first was incapable of numerical evaluation.[2] Cournot further thought that the philosophical probability of inductive con-clusions was proportionate to their 'simplicity.' The idea of

---

[1] For this idea see Couturat, *La Logique de Leibniz d'après des Documents Inédits* (1901), p. 254, *etc.*, and the same author's *Opuscules et Fragments Inédits de Leibniz* (1903), p. 174 and *passim*.

[2] Cf. Fries, *Kritik der Prinzipien der Wahrscheinlichkeitsrechnung*, p. 18. Cf. also Cournot, *Essai sur les fondements de nos connaissances* (1851), Vol. II. p. 386, where the author says: 'La probabilité philosophique répunge tout à fait à une évaluation numérique.'

simplicity he conceived on the pattern of the simplest curve to be traced through a number of given points.[1]

Thus the first authors to deal with Inductive Probability tended—as far as their views are clear enough to permit any definite statement as to their nature—to take what we have called (Chap. IX, §1) a dualistic opinion of the subject. Their dualism, further, appears to have been of a radical type, regarding the nature of the two concepts of probability as *toto coelo* different.

The first attempt at a reconstruction of arguments of Inductive Probability within the frame of a calculus is due to Keynes. The appearance of his *Treatise on Probability* in 1921 can truly be considered a milestone in the history of the Logic of Induction in general and as marking the inception of the Logic of (Real) Inductive Probability in particular.

Keynes dealt chiefly with the Argument from Confirmation. His proof of the Theorem of Confirmation, however, is neither complete nor correct in every detail. Besides it is unnecessarily complicated. Keynes also made a rudimentary attempt to deal with the Argument from Simplicity.

Keynes's work found a continuator in Nicod, who in his *Problème logique de l'induction* (1923) simplified Keynes's proof of the Theorem of Confirmation.

On two points, however, Keynes and Nicod were in disagreement.

The first was, whether continued confirmation of a law will tend to increase its probability towards the maximum value 1 as a limit or not. Keynes answered the question affirmatively and Nicod negatively.

In order that the probability of a law may be increased through confirmation it is necessary that two conditions should be fulfilled (cf. above p. 248), *viz.*, that the probability *a priori* of the law is not minimal, and that the probability of a new confirmation of it relative to the previous confirmations is not maximal.

As a criterion of increase towards maximum probability

[1] Cournot, *Essai sur les fondements de nos connaissances*, Vol. I, p. 82: 'En général, une théorie scientifique quelconque, . . . peut être assimilée à la courbe que l'on trace d'après une définition mathématique, en s'imposant la condition de la faire passer par un certain nombre de points donnés d'avance.'

Keynes and Nicod employ the following: The probability of continued confirmation of the law on the assumption that the law is false approaches minimum probability as its limit. (This criterion can easily be derived from the second of the two equivalent criteria mentioned above on p. 248.) In the subsequent discussion, however, Keynes replaces this criterion by a new condition which, as Nicod acutely observes, is stronger. The new criterion is that, relative to the assumed falsehood of the law *and* its confirmation in n instances, the probability of a new confirmation is not maximal.

In the opinion of Keynes, the increase towards maximum probability is to be secured in virtue of a postulate concerning the constitution of reality. This is the Postulate of Limited Variety. According to it, as I think we must understand it (cf. above p. 136), there is in the Universe of Properties concerned a finite number of logically totally independent ('generating') properties, of which all other properties in the universe are presence-functions. There is thus also a finite number of possible nomic connexions between members of the universe. This finitude of the possibilities, in the opinion of Keynes, would secure not only that any law is not minimally probable *a priori*, but also that the probability of a law's continued confirmation relative to its assumed falsehood tends to zero.

The criticism of the Postulate of Limited Variety is not relevant in this connexion. We have made some comments on it elsewhere. (Cf. Chap. V, §3.)

Nicod seems to agree with Keynes that the Postulate of Limited Variety secures the non-minimal nature of the probabilities *a priori* of the laws. He adds, however, the important observation (cf. above p. 165) that this, strictly speaking, is true only if the postulate is strengthened so as to assert, not simply that the number of 'generating' properties in a certain universe is finite, but that there is a non-minimal probability that it is not greater than a known finite number n.

Nicod criticizes Keynes's view that the postulate would also make a new confirmation of a law non-maximally probable relative to the assumption that the law is false but has been confirmed n times. The nerve of Nicod's reasoning can be stated as follows: A law or Universal Implication $H \subset A$ may be false and yet a proportion 1 of H be included in A. But if this be the case, *i.e.*, if 'practically all' though not all H's are A, then it is reasonable to think the probability to be 1 or maximal that a random thing x which is H is also A. This again, in the opinion of Nicod, would contradict the criterion—both in its original and in its stronger form as used by Keynes—for the increase of the law's probability towards maximum probability through continued confirmation. Hence Keynes would not, by means of his Postulate of Limited Variety, have succeeded in securing the increase in probability towards the maximum value as a limit.

If we accept the reconstruction of the Argument from Confirmation which has been suggested in this book, we must reject Nicod's objection as being beside the point. The probability which, according to Keynes, must not be maximal, is *not* the probability that (on the assumption that the law is false) a *thing* which is H is also A. It is the probability that a *property* which is not a Necessary Condition of H and yet is present in the presence of H in n things, is present in the presence of H in a new thing also. Concerning this probability Nicod's counter-argument can prove nothing. On the other hand, Keynes's postulate cannot prove this probability not to be maximal. What the postulate *can* prove, however, is that the original form of the criterion of increase towards maximum probability is applicable, *i.e.*, that the probability that a property which is not a Necessary Condition of H is present in n things which are instances of H tends to zero as n increases. Keynes's reasoning from the Postulate of Limited Variety would thus have been correct, provided he had not replaced the original criterion for increase towards maximum probability by the stronger criterion.

Actually, as has been shown in Chap. IX, §3, the probability of a law which is continuously confirmed can be proved to approach the maximum value 1 as a limit quite independently of whether the number of properties in the universe is finite or infinite. If the number is infinite, the proof presupposes the Axiom of Continuity which was not used by Keynes. This axiom, however, is not, like the Postulate of Limited Variety, an assumption as to the constitution of reality, but part of the 'implicit definition' of probability embodied in the axioms of the calculus.

The second point on which Keynes and Nicod disagreed was whether confirmation can increase a law's probability without contributing to the elimination of concurrent laws. Keynes answered the question negatively and Nicod affirmatively.

Keynes tried to prove his opinion from the condition that an instance, in order to increase a law's probability, must not itself be maximally probable. He seems to think that if a new instance is maximally probable in relation to the previous instances then, it must be identical with one of them. Conversely, if the instance is not maximally probable relative to the previous instances, then it must differ from any one of them (separately) in at least one property. And this means that the instance will eliminate at least one concurrent law, not eliminated by the previous instances. The proof is as follows:

Take a property of x_1 which is not a property of x_{n+1}. Take a property of x_2, a property of x_3, . . ., and a property of x_n which is not a property of x_{n+1}. The sum of those properties is not a property of x_{n+1}. But it is a property of x_1 and of x_2 and . . . and of x_n. Consequently, x_{n+1} eliminates at least one property from the set of common properties of the first n instances. This means that at least one property has also been eliminated either from a set of initially possible Sufficient Conditions or from a set of initially possible Necessary Conditions of a given conditioned property, according to what kind of nomic connexion we are looking for.

Keynes's proof of the necessity of elimination is erroneous if it assumes that a maximally probable confirmation must be afforded by a thing which is identical with some of the previous things affording confirmations of the law. And unless we accept the highly questionable view known as *identitas indiscernibilium*, the proof is erroneous also in assuming that instances must differ in some property from the same universe as the conditioned property and the possible conditioning properties, in order not to be identical. His argument is correct only in so far as it asserts that a (positive) instance (of the conditioned property) differing from every one of the previous (positive) instances (of the conditioned property) in at least one property (of a certain range of properties) must contribute to the process of elimination.

INDUCTION AND INVERSE PROBABILITY

Nicod tried to prove against Keynes that elimination is not necessary for the increase in probability and that even if two instances were identical they may each contribute to this increase. The nerve of his reasoning is to deny the contention that, if a new instance is identical with a previous one, we could infer and thus determine with maximum probability that it is a confirming instance of the law. By 'identity' he here means agreement in the conditioned property and all the possible conditioning properties, except the one which occurs in the law we are contemplating. On this definition it is clear that we cannot infer from the 'identity' that the new instance will possess the conditioning property which occurs in the law. This, obviously, is correct. But it only proves that, on Nicod's conception of the identity in question, the second instance cannot be inferred from the first, but not that its probability is not maximal. Nicod, moreover, does not mean by the 'identity' of the instances the same thing as was obviously contemplated by Keynes.

If we accept the reconstruction of the Argument from Confirmation which has been suggested in this book, the dispute between Keynes and Nicod as to the necessity of elimination for increase in probability can only be settled on the basis of an *interpretation* of axiomatic probability. On the Frequency Interpretation, we should have to admit with Keynes that increase in probability reflects an underlying process of elimination.

Thus, against the background of our reconstructive examination of the Argument from Confirmation, we must arrive at the somewhat paradoxical conclusion that Keynes upheld the correct views with insufficient or erroneous reasons, whereas Nicod used substantially valid reasoning in support of incorrect views.[1]

[1] In my thesis *The Logical Problem of Induction* (1941) I discussed at some length the Argument from Confirmation and Inductive Probability in general. The discussion, however, was from several points of view defective and contained some errors. I there supported, *inter alia*, Nicod against Keynes in the controversies over maximum probability and over elimination which we have discussed above.

In my publication *Ueber Wahrscheinlichkeit* (1945) are found the essentials of the reconstructive examination of the Argument from Confirmation given in this book. I also dealt there with the Paradox of Confirmation. The completed form of the reconstruction of the Argument from Simplicity and the entire reconstruction of the Argument from Analogy have emerged during the course of my work on the present treatise.

Keynes and Nicod approached the problems of Inductive Probability on what we have here called monistic lines. After their work, the subject remained very little cultivated for about twenty years, but is now receiving fresh attention from a number of logicians and philosophers. The current trend of approach seems on the whole to be of the dualistic type.

The most noteworthy contributions to the study of Inductive Probability in recent years are perhaps those of Carnap. They are, however, so far known to me only from preliminary reports[1]. Carnap distinguishes between two concepts which he calls probability$_1$ and probability$_2$ respectively. The second is a relative frequency. The first is a ratio of possibilities and, as such, related to but not identical with our concept P of axiomatic probability on the Possibility Interpretation (cf. above p. 218). It is probability that is regarded as relevant to induction. We shall not here discuss the details of Carnap's approach. Only the following observation will be made:

Carnap's concept of degree of confirmation (Inductive Probability) is *not* identical with what is called in this inquiry (Real) Inductive Probability. It is *not* probability merely as an attribute of laws, *i.e.*, Universal Implications and Equivalences. Carnap seems to take the view that the degree of confirmation of the law itself, as opposed to its next or n next instances, is bound always to be 0. The law is then being treated as an infinitely long conjunction of its instances. This, no doubt, is a possible attitude. Only I am afraid that a view, on which one cannot discriminate between various degrees of Inductive Probability as a genuine attribute of laws, must fail reconstructively to illuminate a major part of the reasoning which is actually conducted in the field.

In his recent work *Human Knowledge* (1948), Bertrand Russell deals in some detail with problems of induction and probability. Russell also starts from a dichotomy. He distinguishes between what he calls mathematical probability and degree of credibility. The first 'is numerically measurable and satisfies the axioms of the probability-calculus.' It thus corresponds, roughly, to what we have called axiomatic probability. The second is

---

[1] *On Inductive Logic* in Philosophy of Science *12*, 1945. *The two Concepts of Probability* in Philosophy and Phenomenological Research *5*, 1945.

the probability of an individual case on 'all relevant evidence.' It thus corresponds, approximately, to what we have called chance. Credibility in Russell's sense is closely related to the idea of *rational* estimations of probability; Russell also relates it to *equal* possibility, but seems not to be aware of a kinship with *random* distribution.

It is not quite clear to me in what way Russell's degree of credibility would be 'quite a different concept from that of mathematical probability.' Russell thinks that credibility can sometimes be numerically evaluated and thus turned into mathematical probability. He does not decide on the issue whether credibility, which is *not* numerically evaluated, is *non*-numerical in any 'deeper' sense of the word than that it is a probability, on the numerical value of which we do not pass a judgment. (Cf. above p. 173.) There is, so far as I can see, nothing in Russell's view which explicitly contradicts the view which would correspond to that taken in this inquiry, that 'mathematical probability' and 'credibility' have the axiomatic framework in common, and that their difference is not one between two sorts of *probability* but one between two types of *evidence* (for one and the same type of probability).

INDEX OF NAMES

INDEX OF NAMES

SUBJECT INDEX

Printed and bound by CPI Group (UK) Ltd, Croydon, CR0 4YY

01/11/2024

01782615-0003